YAKUBOVA OLTINOY ABDUGANIEVNA

OBSTETRICS AND GYNECOLOGY

" BLEEDING IN OBSTETRICS AND HEMORRHAGIC SHOCK "

ISBN 978-93-5872-977-1

9 789358 729771

Book	:	Obstetrics and Gynecology
Author	:	Yakubova Oltinoy Abduganievna
Publisher	:	Taemeer Publications
Year	:	'2024
Pages	:	102
Title Design	:	*Taemeer Web Design*

Compiler:

O.A. Yakubova - Head of the Department of Obstetrics and Gynecology of the Faculty of Medical Education, Ph.D., associate professor

Reviewers:

G. A. Ikhtiyarova , Ph.D., Professor, Head of the Department of Obstetrics and Gynecology, Faculty of Medicine, Bukhara Medical Institute

D. R. Ph.D. **Khudoyarova , professor,** head of the Department of Obstetrics and Gynecology, Faculty of Medicine, Samarkand Medical Institute

1. NOTES ON CONDITIONAL ABBREVIATIONS
2. AH - arterial hypertension
3. ADG- antidiuretic hormone
4. ABP - arterial blood pressure
5. ABV - circulating blood volume
6. WHO - World Health Organization
7. GBO - hyperbaric oxygenation
8. GCS - glucocorticosteroids
9. OI - oxygen inhalation
10. ABB - acid-base balance
11. CT - computer tomography
12. DICS- Disseminated intravascular coagulation syndrome (DIST)
13. CVP - central venous pressure
14. CNS - central nervous system
15. DNLP displacement of a normally located placenta
16. MRV - minute respiratory volume
17. NB - number of breaths
18. PAV - pulmonary artificial ventilation
19. SHI- shock index
20. UE- ultrasound examination
21. ARF - acute renal failure
22. ALF - acute liver failure
23. ARF - acute respiratory failure
24. AHF - acute heart failure
25. FFP - fresh frozen plasma
26.

ANNOTATION

The study manual is a study manual prepared for treatment and professional education students and clinical supervisors of medical universities. For students of medical institutes, clinical residents and magistracies, as well as special training doctors, it is set as a priority task to radically increase the speed, technological and efficiency of emergency medical care for bleeding in obstetrics. Uzbek medicine has a very important task in front of it - it is to learn to manage emergency medical conditions and continuously follow this path. Practical skills, algorithms, and a glossary of key terms used in emergency practice are also presented. The training manual can also be used by emergency and non-urgent medical teams, resuscitation and intensive care units. The manual is written based on the model curriculum approved by the Ministry of Higher and Secondary Special Education of the Republic of Uzbekistan

CONTENTS
CHAPTER I.

II CHAPTER.

CHAPTER III.

INTRODUCTION

In the years of independence, in a broad sense, the work to eliminate medical illiteracy began to be implemented as an important issue within the state. It is not an exaggeration, today exactly such a task also applies to urgent medical care. One of the most important areas of reform is the formation of emergency medical service, especially if it depends on the fate of mother and child. The modern approach to the organization of emergency care at various stages, diagnostic and treatment methods, personnel training and education differs in many ways compared to 15-20 years ago. By itself, the demands for emergency medical service workers in obstetrics, all obstetrician-gynecologists and all medical staff have increased. However, in most developed countries, pre-hospital first aid is the responsibility of emergency medical personnel and paramedics. Despite the existence of educational publications on urgent medical problems, today there is a great need for educational textbooks for medical personnel that cover emergency situations, their diagnosis and treatment methods. This tutorial is intended to meet that need to some extent. Special attention is paid to the mechanisms of its origin and the necessary diagnostic and treatment algorithms in obstetric bleeding emergencies. In more than 10 decisions and decrees issued in the first six months of 2017, the Honorable President emphasized the "deplorable condition of the ambulance service and the need to immediately regulate this system" and gave clear instructions to eliminate them. In particular, the President of the Republic of Uzbekistan Sh. In Mirziyoev's report, "Critical analysis, strict order, discipline and personal responsibility should be the daily rules of every leader's activity", 4 out of 6 (80%) of the 10 deficiencies in the medical system are directly related to the emergency medical care system. Radically strengthening the speed, technological and efficiency of emergency medical care for bleeding in obstetrics has been defined as a priority task. Uzbek medicine faces a very important task - it is to learn to manage emergency medical conditions of bleeding in obstetrics and continuously follow this path. Practical skills, algorithms, and a glossary of key terms used in the practice of obstetric emergencies related to hemorrhagic shock are also presented. The training manual can also be used by emergency and non-urgent medical teams, resuscitation and intensive care units.

CHAPTER. BLEEDING IN THE FIRST HALF OF PREGNANCY

1.1. Bleeding in the first half of pregnancy is pregnancy-related bleeding that occurs before the 22nd week of pregnancy, which is due to the inability to carry or carry the pregnancy to term. Failure to carry pregnancy to term includes concepts such as: spontaneous abortion in early and late pregnancy, premature births, and undeveloped pregnancies.

Physiological changes in the ovary during pregnancy. With the onset of pregnancy, regular processes in the ovary stop. The corpus luteum functions in one of the ovaries, the hormones it releases (progesterone and estrogen) create conditions for normal pregnancy.

Estrogens helps to accumulate contractile proteins (actin and myosin) in the uterine muscles, creates conditions for increasing the reserve of phosphorus compounds, which leads to the expansion of blood vessels, ensures the use of carbohydrates by the uterus.

Progesterone causing decidual changes in the endometrium, providing the growth and development of the myometrium, the increase of blood vessels (vascularization) and creating conditions for the implantation of the fertilized egg cell. Progesterone neutralizes the effect of oxytocin, reduces the excitability of the uterus, activates the growth and development of mammary glands.

In order for the pregnancy to end successfully, it is necessary to create conditions aimed at the recognition of fetal antigens by the mother's immune system and the preservation of the pregnancy. In a normal pregnancy, there are progesterone receptors in peripheral blood lymphocytes, and the number of cells in it increases as the gestation period increases. This process is carried out under the influence of antigens of the embryo, the activity of lymphocytes increases and progesterone receptors appear in their composition.

In normal miscarriages, threatened miscarriages and premature births, the percentage of progesterone receptor cells in the second and third trimesters of pregnancy is much lower than in healthy women. In the presence of progesterone or dydrogesterone (duphaston), these lymphocytes (PPBF) produce progesterone-inducing blocking factor mediator protein. A decrease in PPBF has been reported at risk of miscarriage or early labor.

It is known that the rise of rectal temperature to 37 degrees and more is a sign of pregnancy, indicating that progesterone is sufficient. (T.D. Travyanko, Ya.P. Solsky). This sign is important up to 12 weeks of pregnancy, after this period this sign does not serve as information.

Progesterone deficiency in the early stages of pregnancy (up to 12 weeks) basal temperature is 36.5-36.9°C, this sign is the basis for the conclusion about progesterone deficiency and proves that the woman should be prescribed progesterone.

One of the most common complications of pregnancy, which leads to complications during pregnancy and the postpartum period, is the inability to carry a pregnancy. According to various researchers, its occurrence ranges from 5 to 48%. In Uzbekistan, miscarriages in the first trimester are 57.7%, in the second trimester - 17.2%, premature births are 25.1%.

Inability to bear pregnancy in 64 - 74% of cases is caused by hormonal deficiency of ovaries and placenta (Kosheleva N.G. and others; Serov V.N., 2017). Usually, ovarian failure is understood as a decrease in their hormonal activity.

1.2. Spontaneous abortion is the spontaneous termination of pregnancy when there is a chance for the fetus to develop outside the mother's womb. According to the latest data, 7 out of 7 clinically diagnosed pregnancies end in abortion in the first 14 weeks, and more than half of the fetuses have chromosomal defects.

Spontaneous abortion is one of the most frequent obstetric pathologies in medical practice. There are few accurate statistics on spontaneous abortions: their number is considered based on relative data (the ratio of the number of women admitted to the hospital with a spontaneous abortion to the total number of women admitted during the same period). According to various authors, spontaneous abortions account for 2-8% of the total number of pregnancies.

There are many reasons for spontaneous abortions . Factors leading to and causing abortion are distinguished among them. Contributing factors include diseases during pregnancy, poisoning, functional disorders , physical and mental injuries that directly lead to abortion . I. of the process that brings to an end the reasons that lead to miscarriage. L. Braude called "decisive" factors. Such a division of etiological factors can be assumed, because the factors leading to and resolving spontaneous abortion have a joint effect. But this classification does not take into account the genetic defects that cause the development of the fetus to fail and fall.

Spontaneous abortion is the loss of a fetus before the formation of the viability of the fetus (22 weeks of pregnancy).

Spontaneous miscarriages include those in the rhythm

- Threatened abortion (pregnancies can be saved);
- Abortion on the way (the pregnancy goes to the stage of incomplete/complete abortion without being able to continue);
- Premature or incorrect abortion (fetal particles are partially separated);
- Complete abortion (fetal parts are completely separated).

Artificial abortion terminates pregnancy using various means until the fetus is viable.

Criminal abortion is a procedure performed by unqualified persons under conditions that do not meet minimum medical standards.

Septic abortion abortion complicated by infection. Sepsis can be the result of the infection rising from the lower floor of the genital tract after a spontaneous or criminal attack. Sepsis is more likely to develop when fetal particles are retained in the uterine cavity and their removal is delayed. Sepsis is one of the most common complications of criminal abortion when performed with non-sterile instruments.

Treatment. <u>If it is suspected that a woman has had a criminal abortion, her</u> It is necessary to check for infection , at the same time, to rule out injury to the uterus and vagina , thoroughly wash the vagina from any herbs, locally used drugs and caustic solutions .

In threatened abortion usually there is no need for medicinal treatment . Recommend to refrain from sexual intercourse, activities that require different strength, bed routine is not necessary.

If the bleeding has stopped, continue the follow-up at the gynecologist's office.

If bleeding recurs, reassess the woman's condition.

If bleeding continues, assess for fetal viability (pregnancy test/ultrasound) or possible ectopic pregnancy (ultrasound). Continued bleeding, especially if the size of the uterus is larger than expected, can indicate a twin pregnancy or miscarriage.

Do not give hormonal drugs (eg, estrogens or progestins), or tocolytic agents (eg, salbutamol or indomethacin), because they do not stop the miscarriage.

Abortion on the way.

If pregnancy is less than 16 weeks, plan to clean the uterine cavity If this is not possible, promptly : ergometrine 0.2 mg intramuscularly (repeat in 15 minutes if necessary) OR misoprostol (Cytotec) 400 mcg orally (repeat in 4 hours if necessary);

Be as ready as possible to empty the uterine cavity .

If the pregnancy period is more than 16 weeks :

- wait for the expulsion of the fetal egg (fall), then remove the remnants of the fetus by aspiration of the uterine cavity;

- if necessary, inject oxytocin 40 ED in 1l solution intravenously (in saline solution or lactate - Ringer solution) at a rate of 40 drops in 1 minute to help the expulsion of the fetal egg.

- Provide follow-up and control to the woman after the therapy.

A premature abortion

If the bleeding is light to moderate and the pregnancy is less than 16 weeks, use your fingers or a glass forceps to remove the fetal parts from the uterine cavity to the cervix.

Clean the uterine cavity if the bleeding is heavy and the pregnancy is less than 16 weeks.

- manual vacuum aspiration is the most convenient method for evacuation. If it is not possible to carry out manual vacuum aspiration, then evacuation can be performed using a sharp curette.

- if it is not possible to quickly empty the uterine cavity, order ergometrine 0.2 mg intramuscularly (repeat in 15 minutes if necessary) OR misoprostol (Cytotec) 400 mcg orally (repeat after another 4 hours if necessary)

If the gestation period is greater than 16 weeks

- until the expulsion of the fetal egg, inject oxytocin 40 ED in 1l solution intravenously (in physiological solution or lactate-Ringer solution) at a rate of 40 drops per 1 minute.

- if necessary, misoprostol (cytotec) 200 µg vaginally after 4 hours until the expulsion of the fetal egg, but the total dose should not exceed 800 µg;

- remove all the carried fetal particles from the uterine cavity

- Provide follow-up and control to the woman after the therapy.

Complete abortion

- there is no need to evacuate the tissues inside the uterine cavity.

- monitor the woman in order not to miss heavy bleeding

- ensure the observation of the woman

Follow-up of women who have had an abortion

Before leaving the hospital, inform the woman who has experienced a spontaneous abortion: spontaneous abortions are common and 15% terminate spontaneously (one in seven women). Also, reassure the woman that the chances of a successful subsequent pregnancy are high, if the woman does not have sepsis, or if the cause of the abortion is not determined, it can affect subsequent pregnancies (this is rarely observed).

Some women still want to get pregnant after a botched abortion. Convince the woman to postpone the next pregnancy until she is fully recovered.

Counseling is very important for women who are victims of crime. If pregnancy is unwanted, family planning measures related to pregnancy prevention should be taken by women at home immediately (within 7 days).

- if there are no serious complications that do not require further treatment ;

- after receiving the necessary advice on the most convenient method of family planning .

PATHOGENESIS OF TERMINATION IN THE EARLY TERMS OF PREGNANCY

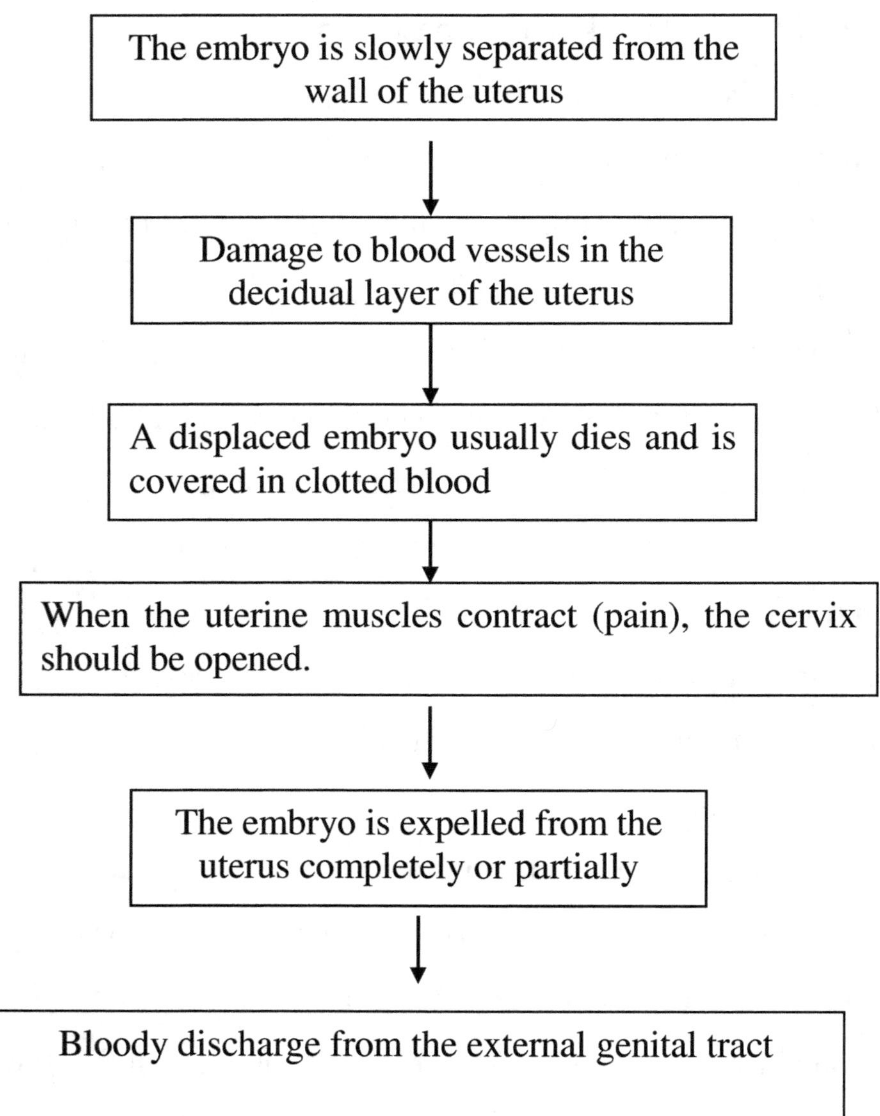

1. 3. Causes of bleeding in the first half of pregnancy

- Infantilism
- Neuroendocrine disorders
- Artificial removal of pregnancy
- Isthmic-cervical insufficiency

- Agreements

- Chromium somal and genetic anomalies

- Defect of egg cells and spermatozoa

- Infectious diseases during pregnancy

- Izoantigen imbalance of maternal and fetal blood

- Diseases and developmental defects of genital organs

- Poisoning of the organism

- Effects of ionizing radiation

- Eating disorder

- Physical injuries

Infantilism is one of the most common causes of miscarriage. In infantilism , the first and subsequent pregnancies are usually terminated by themselves (normal miscarriage). The causes of miscarriage in infantilism depend on: ovaries characteristic of infantilism functional deficiency of endocrine function , insufficient decidual response of uterine mucosa , insufficient hyperplasia of infantile uterine muscles and other necessary processes for pregnancy development unimproved ; in infantilism, there is often an increase in the excitability of the uterus.

Neuroendocrine disorders , including diseases of the internal secretion glands, cause pregnancy to fail in most cases . Termination of pregnancy is often caused by dysfunction of the ovaries, adrenal gland (hyperandrogenism with adrenal gland genesis), hypothyroidism, diabetes and other pathologies. Spontaneous miscarriage is also observed in disorders of the endocrine glands, whose function is not well known. In endocrine disorders, there is no change in the activity of the endocrine glands necessary for the normal course of pregnancy and the development of the fetus.

Endocrine disorders are usually preceded by disturbances in the control of the nervous system, especially its vegetative department. Because of this, vascular disorders and shifts in the acetylcholine-cholinesterase system under the predominance of the acetylcholine indicator often occur. Acetylcholine increases uterine contractions, which leads to miscarriage.

Emotional disorders and other types of neuroses often appear in women who suffer from miscarriage. These changes occur secondary to subsequent disruptions in the neural control of processes that promote pregnancy.

Artificial termination of pregnancy often causes spontaneous abortion later. Especially the artificial termination of the first pregnancy has bad consequences. A number of reasons allow the occurrence of spontaneous abortions after artificial removal of the fetus. Disturbances in the endocrine and nervous system, chronic endometritis and other inflammatory diseases that often appear after abortion are of great importance. When the cervix is dilated and the fetal egg is removed with instruments, there is a possibility of damage to the muscle elements around the internal neck of the uterus and in the isthmic area, which can lead to insufficiency of the cervical canal.

Isthmic-cervical insufficiency occurs as a result of damage to the isthmic part of the cervix during obstetric operations or artificial abortion. As a result of the injury, the muscles of the circulation lose their ability to contract, and the internal neck of the cervical canal remains dilated. In the absence of the cervix, the mechanism that holds the fetal egg in the uterine cavity (narrowed area in the area of the inner throat and neck) is lost. Without sufficient support in the lower part of the uterus, the fetus slowly descends through the internal horn, the lower end of which opens into the cervical canal. Later, uterine contractions occur and the fetus falls out. More late miscarriages occur in cervical insufficiency. Deficiency in isthmic-cervical insufficiency occurs not only as a result of injury (organic insufficiency), but also endocrine disorders, especially in the decrease of the hormonal function of the ovaries (functional insufficiency).

Miscarriage can be caused by adhesions (synechiae), which cause sharp deformation of the uterine cavity after rough destruction, severe types of endometritis (especially endometrial tuberculosis), as well as deep tears of the cervix and cervical - vaginal fistulas.

Chromosomal and genetic anomalies contrary to the development of the fetus lead to its death and miscarriage. Various chromosomal aberrations - aneuploidy, trisomy, monosomy, mosaicism, etc. were found when fetal egg particles were examined. Chromosomal and genetic abnormalities often lead to early miscarriage. The role and occurrence of chromosomal abnormalities in miscarriage continues to be studied. It is estimated that 20% of miscarriages are associated with chromosomal and genetic abnormalities.

Defective changes in the egg cell and spermatozoa before fertilization can be the cause of fetal development and subsequent abortion. Defects appear under the influence of genetic or environmental factors (severe diseases, alcoholism, poisoning, etc.). In such cases, the death of the fetal egg can occur in the early stages of development (during implantation, in the initial period of organogenesis).

Infectious diseases during pregnancy often lead to spontaneous abortions. Acute infectious diseases that appear in the early stages of pregnancy often cause miscarriage. Among infectious diseases, the most common flu plays a major role. Termination of pregnancy is often caused by viral hepatitis, rubella, acute rheumatism of the joints, as well as rare infectious diseases: malaria, typhus, whooping cough, etc. Abortion can be observed in angina, zotiljam, pyelonephritis, cecum and other acute inflammations in various systems and organs.

Circumstances that lead to termination of pregnancy in acute infectious diseases: hyperthermia, intoxication, lack of oxygen, hypovitaminosis and other disorders. Dystrophic processes and hemorrhages appear in the decidual layer; the chorion loses its barrier properties and causative microbes can enter the organs of the fetus.

Chronic infectious diseases can cause termination of pregnancy. Termination of pregnancy in toxoplasmosis, pulmonary tuberculosis, brucellosis, ulcer, chronic malaria is less common than in other acute infectious diseases. Fetal miscarriage can be observed in organic diseases of the heart with blood circulation disorders, chronic glomerulonephritis and severe hypothyroidism and other systemic diseases.

Isoantigen imbalance of maternal and fetal blood causes miscarriage if the blood of the husband and wife inherits antigens of the father according to the Rhesus-factor and AVO system and other Rhesus antigens. Antigens of the fetus against maternal antigens enter the body of the pregnant woman through the placenta and lead to the formation of specific antibodies. Antibodies to the Rhesus factor, blood group and other antigens enter the fetus through the placenta, causing hemolytic disease and other disorders that cause fetal death and miscarriage. Often, not the first pregnancy, but repeated miscarriages occur, because after multiple pregnancies, the body's sensitization increases.

Diseases of the genitals and developmental defects can also cause spontaneous abortion. If inflammatory diseases of the genitals are accompanied by changes in the function and structure of the main layers of the endometrium and myometrium, they can cause miscarriage. Adhesions, synechiae, fixed retroflexion, pelvic tumors, and other processes that impede the growth of the pregnant uterus can be the cause of abortion.

In uterine myoma, the fetus may drop out if the graft is in the area of the mucous membrane covering the myomatous node, it becomes thinner and cannot provide the fetus with normal nutrition and development, and as a result, it falls out.

Miscarriages are often observed in cases of uterine malformations. In case of two-kingdom, two-kingdom, one-kingdom uterus and other developmental anomalies, pregnancy can continue until the physiological end, but in most cases it is terminated early (miscarriage, premature births). Endocrine disorders accompanying pregnancy termination, decidual reaction and other processes typical for normal pregnancy create conditions.

Body intoxication (especially chronic) often leads to the death and miscarriage of the fetus. Lead, mercury, gasoline, alcohol, nicotine, aniline compounds and other toxic chemicals are of great importance.

Ionizing radiation occupies a special place among the causes of pregnancy termination. The fetus is highly sensitive to the effects of radiation, its damage occurs at a dose that an adult body can handle. In this regard, pregnant women are prohibited from working with a radiation source; Pregnant women are also prohibited from working in various branches of chemical production.

Eating disorders can also cause miscarriage. Spontaneous abortion can occur during starvation, especially if it is at the level of alimentary dystrophy. Termination of pregnancy can be caused not only by limiting general nutrition, but also by poor quality nutrition. Deficiency of vitamins, which are of great importance in the development of the fetus and participate in the management of physiological processes during pregnancy, plays a special role. Experimental data show that the development of the fetal egg (fetus and membranes) and miscarriage is caused by a lack of vitamins E, A, C, B, etc. in food. In hypovitaminosis, hemorrhages and dystrophic processes appear in the decidual layer and chorion, which lead to the death of the fetus.

The above can be said to conditionally refer to the factors that cause miscarriage. It should be mentioned that these reasons are not only the cause, but can directly cause miscarriage. However, in addition to the above-mentioned etiological (causing) factors, the probability of abortion increases under the influence of additional factors: physical and mental injuries.

Against the background of infantilism, inflammatory diseases, endocrine disorders, poisoning and other factors, physical injuries (blows, broken bones, body injuries, etc.), excessive strain (lifting and carrying heavy objects) can make the fetus abort. can appear after a strong mental shock.

In healthy women, pregnancy is preserved even with severe injuries (blows, broken bones of the pelvis and legs) and unconscious effects on the nervous system.

In abortion, the fetal egg slowly separates from the uterine wall; in which the blood vessels of the decidual layer of the uterus are damaged and bleeding occurs, the strength of which depends on how far the fetus has moved and the number of damaged blood vessels. A displaced fetal egg usually dies and is surrounded by or absorbed by the shed blood. Under the influence of contraction of the uterine muscles (pain), the opening of the cervix and the complete or partial expulsion of the fetal egg occurs. If the fetal egg is born single (together with the fetal membranes and amniotic fluid), it is said to be a spontaneous miscarriage. If the fetal egg breaks out of the uterine cavity, it is called a double miscarriage.

A single miscarriage is less common than a double miscarriage, especially in the early stages of pregnancy. In a double miscarriage, the membranes and parts of the placenta are usually trapped in the uterus. Due to the contraction of the uterus, a woman feels painful pains in the lower abdomen, and their strength is greater in late fetal dreams. In the first abortions, the sensation of pain will not be very strong or may not be present at all. The feeling of pain is not so strong in a miscarriage associated with cervical insufficiency. Abortion is usually observed together with bleeding, and its strength depends on the stage of this process and the duration of pregnancy.

A late miscarriage is similar to labor: the cervix develops and opens, the amniotic fluid passes, the fetus is born, and then the placenta descends. Blood loss is often observed if there is a violation of the migration of the placenta and expulsion from the uterus; in these cases it can be large.

Certain risk factors are important for each trimester of pregnancy.

in the first trimester can be caused by:

- genetic factors;

- endocrine: ovarian dysfunction, diabetes mellitus , diseases of the thyroid and adrenal glands;

- developmental defects of the uterus;

- somatic and infectious diseases of the mother, especially latent mycoplasma, chlamydia, viral, toxoplasma, ureaplasma and bacterial infections of the urinary tract.

in the II trimester is of particular importance: cervical insufficiency, uterine fibroids and developmental defects, placental insufficiency, as well as infectious diseases and somatic pathology of the mother.

1. 4. Examination of women who could not carry the pregnancy to term

It is necessary to start the examination of women with a history of premature births and premature births, and it is necessary to carry out the following:

- examination by a therapist for the purpose of identifying chronic infection foci, their sanitation, and treatment of identified somatic pathology;

- examination at the medical -genetic center ;

- examination of ovaries and other glands of internal secretion (functional diagnostic tests and the amount of hormones in the blood);

- during cervico- and hysterosalpingography , to assess developmental defects of the uterus, cervical insufficiency, the inner horn of the uterus (in vaginal examination and ultrasound examination) and to rule out uterine tumors ;

- to determine the duration of pregnancy, placental characteristics, uterine tone, fetal developmental defects during UE of the pelvic organs;

- detection of urogenital infection and its subsequent treatment .

- examination of hormone excretion (estrogens, progesterone, chorionic gonadotropin, placental lactogen, 17-ketosteroids);

- colpocytological examination, control of basal temperature in the early stages of pregnancy;

- bacterioscopic and bacteriological examinations of the discharge of the mouth-nasal cavity, urethra, cervix and urine;

- examination to detect hidden infection (chlamydia, ureaplasmosis, mycoplasmosis, etc.).

1. 5. Treatment measures in threatened abortion

The treatment of threatened abortion is complex, etiopathogenetic, taking into account the individual characteristics of the female organism, and it should include the following:

1. Bed routine, pregnant woman's desired diet.

2. Sedative therapy (purserum tincture - 20-30 drops 3 times a day, valerian extract - 2 tablets 2 times a day, novopassit - 5-10 ml 2-3 times a day, diazepam - 2 times a day, phenazepam 0,002 r2 -0,0005 rtimes a day or - 2-3 tablets before sleep).

3. Hormonal therapy with progestins or their combination with estrogens and chorionic gonadotropin. At the age of ovarian hypofunction and genital infantilism, hormonal treatment should be started from the 5th week of pregnancy; it can be

continued until 27-28 weeks of pregnancy , depending on the functional activity of the placenta and the effectiveness of the treatment.

Drugs used:

- progesterone (intramuscular 1 ml of 1% solution once a day for 10 days under the control of colpocytological data); in case of progesterone deficiency, 1 ml of 12.5% solution is applied intramuscularly once a week (not recommended due to the effect on the fetus);

- duphaston - 4 tablets at the same time (40 mg), then 10 mg 2-3 times a day with the continuation of treatment for 1 week after the symptoms of the risk of miscarriage disappear . The dose of the drug is gradually reduced;

- ethinyl estradiol (microfolin) 1 tablet a day or 2 tablets 2 times a day until 12-14 weeks of pregnancy (if vaginal discharge is observed, it is given together with progestins);

- chorionic gonadotropin (prophase) intramuscularly in 500-1000 ED units 2 times a week until 12-14 weeks of pregnancy (indicated by low excretion of chorionic gonadotropin).

4. Termination of uterine contractions (tocolytic therapy), aimed at:

- starting from the 16th week of pregnancy, β -adrenomimetics are used to activate uterine β -receptors to increase myometrial tone (ginipral, brikanil, partusisten, etc.). Ginipral is often used (0,0005 rfrom 3-4 times a day or by infusion of 2 ml of 0.005% solution in 500 ml of isotonic sodium chloride solution at a rate of 10-12 drops per 1 minute);

- to reduce the myogenic tone of the uterus (no-shpa 1-2 tablets 2-3 times a day or 2-4 ml of 2% solution intramuscularly or intravenously; papaverine hydrochloride 2 ml of 2% solution intramuscularly 2-3 times; baralgin 1- 2 tablets 2-3 times a day or 5 ml intramuscularly 1-2 times a day; magnesium sulfate 10 ml 25% solution intramuscularly 2 times a day or intravenously 10 ml in 300 ml 5% glucose solution);

- reduction of myometrial tone due to blockade of m-cholinerceptors of the uterus (methacin 1 ml of 0.1% solution intravenously 2 times a day or 2 ml in isotonic solution in 400 ml; platyfillin hydrotartrate 1 ml of 0.2% solution intramuscularly 2-3 times a day) ;

- Reduction of uterine tone due to inhibition of prostaglandin synthetase (diclofenac-sodium 0,05 ror 3 ml of 2.5% solution intramuscularly 1 time a day; indomethacin 0,025 r3-4 times; acetylsalicylic acid 0,25 r2 times a day).

5. In order to improve metabolic processes in the feto-placental complex :

- ascorbic acid (intravenous 3-5 ml of 5% solution in 100 ml of isotonic solution);

-tocopherol acetate (vitamin E) (1 capsule 1-2 times a day or 1 ml of 10% solution intramuscularly 2 times a day);

- polyvitamin drugs .

6. Antibacterial therapy:

- ampicillin (1 r intramuscularly 4 times a day for 7 days),

- cefazolin (1 r intramuscularly 2-4 times a day for 5-7 days).

lamidiosis , mycoplasmosis or ureaplasmosis is detected, treatment must start from the 14th week of pregnancy . In chlamydia, rulid 150 mg 2 times a day or rovamycin 3,000,000 ED units 3 times a day for 10 days, in mycoplasmosis and ureaplasmosis - macropen 400 mg 3 times a day for 10 days

7. In surgical correction of cervical insufficiency:

- placing a circular stitch on the cervix (pregnancy maintenance therapy is considered to be the most effective at 13-17 weeks of pregnancy) at 37-38 weeks;

- insertion of an obstetric pessary.

1.6. Infected abortions

Infected abortion often occurs as a result of the introduction of causative microbes into the uterus during criminal interventions, in rare cases, it can develop through a hematogenous or lymphogenous route. Complications of artificial abortion are also possible if this intervention is performed when there are contraindications (inflammatory cysts in the genitals, infectious diseases of extragenital origin).

The purpose of examining the patient is to assess the spread of infection and identify the causative agent.

Depending on the spread of infection, the following are distinguished :

- uncomplicated infected abortion, in which the infection is located in the uterus, limited to the fetal egg and the decidual layer of the uterus ;

- complicated infected abortion, in which the infection has spread outside the uterus, but is limited to the small pelvic area (myometrium, uterine tubes , blood vessels, ovaries , parametral cell and pelvic peritoneum are injured);

- septic abortion, in which the infection spread beyond the small pelvis and became a generalized infection (sepsis with invisible metastases - septicemia, sepsis with metastasis - septicopemia, peritonitis, acute or spreading thrombophlebitis, septic shock, anaerobic sepsis).

Table 1

Criteria for determining an infected abortion i

Determination criteria are i	Infected abortion		
	uncomplicated	complicated	septic
Complaints	Subfebrile temperature, the body temperature increases, but there is no chill , pain in the lower abdomen , pathological secretions	Body temperature 38 °C, chills , pain in the lower abdomen, pathological secretions	Body temperature higher than 38 °C, hectic type , chills with shivering, pain in the lower abdomen, pathological discharge
The intensity of toxicity	weak	Medium strength	Sharp strong
General condition	Satisfactory	Medium weight	Heavy
Hemodynamic	Tachycardia.	Tachycardia, BP	Tachycardia,

indicators i	KB norm al	changes	bradycardia, decreased BP
Diuresis	norm a	It can decrease	Oligoanuria
Clinical analysis of blood	moderate leukocytosis and anemia	High leukocytosis, leukocyte formula shift to the left , SOE (erythrocyte sedimentation rate) accelerated , hemoglobin decrease	Increased leukocytosis or leukopenia, left shift of the leukocyte formula, signs of toxic anemia, toxic granularity of neutrophils, high leukocyte count
Biochemical analyzes of blood	Without change	Hypo- and dysproteinemia, hypercoagulation	Strong changes , coagulopathy
The degree of injury to the pelvic organs	Endocervicitis, colpitis, endometritis	Endocervicitis, colpitis, metro-endometritis, salpingoophoritis, parametritis, pelvioperitonitis	Endocervicitis. colpitis, metroendometritis, metrothrombophlebitis , salpingoophoritis, parametritis, pelvioperitonitis, peritonitis

<u>Doctor tactics</u> . Uncomplicated infection against the background of antibacterial and infusion therapy in abortion they remove the fetal egg (active method) or they don't do this operation until the infection is suppressed (monitoring tactics).

Table 2

Treatment of infected abortions

Treatment measures	Medicines
Antibacterial therapy	Intravenous delivery of broad-spectrum antibiotics in massive doses : cephalosporins (claforan, kefazol, cefamezin po 6 8 гv sutki), augmentin, tienam, meronem,

	they are used together with aminoglycosides - gentamicin (0,5 г from 3 times a day between the muscles), amikacin (0,5 гfrom 3 times a day), and at the same time metragil is sent intravenously (100 ml 2-3 times a day)
Infusion therapy (2 лmore than	Protein preparations albumin, dry or native plasma, protein. Reopoliglyukin, polyglyukin, polyion solutions . Hemotransfusion - according to the indication.
Drug therapy	Dimedrol, tavegil, pipolfen, tranquilizers, cardio-tonic substances, prednisolone, hydrocortisone, vitamins of group C and V, unitiol, protease inhibitors (trazilol, gordox), hepatoprotectors (essentiale, karsil)
Surgical intervention	In complicated and uncomplicated abortion, cleaning of the uterine cavity from the remnants of the infected fetal egg , extirpation of the uterus with uterine tubes - in septic abortion

Complicated infected abortion requires observation-active tactics: initial antibacterial and detoxification therapy; after the inflammatory process subsides - cleaning the uterine cavity from infected fetal remains, administering hormonal therapy, antibiotics and infusion solutions.

In septic abortion, complex antibacterial and detoxification therapy is carried out, operative treatment is aimed at eliminating the source of infection - extirpation of the uterus with uterine tubes and drainage of the abdominal cavity. Multi-component intensive therapy is carried out in the postoperative period.

Intensive therapy is carried out until the clinical recovery of the woman and the normalization of laboratory parameters.

The treatment of infected abortions is complex, aimed at eliminating the causative agent, restoring metabolic disorders, and eliminating the source of infection.

1.7. Cervical pregnancy

The etiology of cervical pregnancy has not yet been determined. Pregnancy in the cervix, more precisely in the transition from the cervix to the uterus (isthmic-cervical), placenta cervicalis, s. Isthmicocervicalis pregnancy with placenta is understood as such a pathology when, depending on its location, the placenta adheres to the transition between the cervix and the body of the uterus (lower segment) and the cervix. In this case, the chorionic villi penetrate into the muscle tissue of the cervix, sometimes even into the parametrium. Similar pathology is rare and is recorded in case reports; but it is extremely dangerous because the bleeding is profuse and leads to death. Special attention is paid to dystrophic changes of the endometrium after abortion.

The clinical picture is characterized by major bleeding, bleeding occurs in the first or second half of pregnancy. It is caused by displacement of the placenta or rupture of thinned blood vessels in the cervix. In most cases, the pregnancy is interrupted in the first half of the pregnancy, in rare cases it can reach the term.

Diagnosis can be a serious challenge . Among the signs of pregnancy on the cervix: the place where the placenta is located, as if the cervix is swollen, or the place where the entire fetal egg is located is swollen , there is an external cervical eccentricity . Sometimes the cervix looks like an egg-shaped tumor, the bottom of which is located with thinned edges. When viewed through the vagina, the size of the cervix is much larger than the body of the uterus, and it can be mistaken for a fibroid node under the peritoneum.

It is necessary to compare cervical pregnancy (in the early stages of pregnancy) with abortion on the way. Their common symptom is heavy bleeding from the uterus. But in a miscarriage, the uterus is egg-shaped, and in a cervical pregnancy, it is asymmetrical, because one wall of the cervix is bulging.

It is necessary to compare cervical pregnancy with uterine fibroids. Its correct identification is based only on the anamnesis (menstruation, other symptoms of pregnancy), eccentric location of the cervix. In the second half of pregnancy, it is necessary to deny that the placenta lies in front of the bleeding; against it, the placenta is found in the cervical canal, the uterus is located close to the internal throat, and the cervix moves eccentrically.

Treatment is uterine extirpation without removing the embryo through the vagina. Incorrect intervention through the vagina - scraping the uterine cavity or separating the placenta with the fingers - leads to profuse bleeding, often ending in death. Similar bleeding is observed in ineffective separation of the placenta due to the opening of large blood vessels. No therapeutic measures (tamponade, sutures) can stop the bleeding, even in the case of immediate uterine extirpation, it is impossible to save a woman who has lost blood from death.

The consequence is very serious, because timely detection of pregnancy in the cervix and its treatment will lead to an error.

1.8. Diagnosis of vaginal bleeding in the early stages of pregnancy

Ever-present symptoms	Symptoms and signs that sometimes appear	Probable diagnosis
Light bleeding (takes more than 5 minutes for a clean diaper or underwear to clot) Close the cervix The size of the uterus is suitable for the duration of pregnancy	Cramps / pains in the lower abdomen The uterus is softer than normal	Threatened abortion
Light bleeding Abdominal pain Close the cervix The uterus is a little big The uterus is a little soft	• Fainting • Painful structure in the area of excess • Amenorrhea • Pain when moving the cervix	Ectopic pregnancy
Light bleeding Close the cervix	• Mild cramping / pain in the lower abdomen	Complete abortion

The uterus is slightly smaller than the gestation period The uterus is a little soft	• The release of fetal parts during separations in a namnez	
Heavy bleeding (takes less than 5 minutes for a clean diaper or underwear to clot) The cervix is open The uterus is suitable for the duration of pregnancy	• convulsions / pains in the lower abdomen • B is very painful • fetal fragments do not come out in separations	ort on the road
Heavy bleeding The cervix is open The uterus is suitable for the duration of pregnancy	•convulsions / pains in the lower abdomen •B is very painful •fetal fragments do not come out in separations	N complete abortion
Heavy bleeding The cervix is open The uterus is larger than the gestation period The uterus is softer than usual the release of fetal fragments in fragments , reminiscent of grape shingles	•Nausea/vomiting •Spontaneous abortion •convulsions / pains in the lower abdomen •Ovarian cyst / easy to tear •Early onset of preeclampsia •Absence of fetus	El Bugoz

Table .

1. 9. Identification and treatment of abortion complications

Symptoms and signs	complications	treatment

• Pain in the lower abdomen • Pain i rradiation s i • Painful uterus • Prolonged bleeding • Fatigue • High temperature • Foul-smelling vaginal discharge • Cervical discharge purulent • B is painful when the cervix is moved	Infection/sepsis	Antibiotics should be started as early as possible, before the age of manual vacuum aspiration
• Pain in the lower abdomen • Pain i rradiation s i • Abdominal swelling • Abdominal muscle tension (tight or tight stomach) • Pain in the shoulders • Nausea / vomiting • High body temperature	Injury to the uterus, vagina and intestines	a laparotomy to repair the injuries , and at the same time manual vacuum perform the aspirating action . Ask for help if necessary .

1.10. Ectopic pregnancy

An ectopic pregnancy develops when the embryo implants outside the uterus . Fallopian tubes are the most common place of ectopic implantation (90% more).

Depending on the location and rupture of the fetal egg, the symptoms and signs differ greatly from each other. Culdocentesis (rectal puncture of the uterine cavity (posterior dome)) is an important method of detecting an interrupted pregnancy, but is less commonly used than serum pregnancy tests and UE. If non-coagulable blood is detected, start prompt treatment.

Symptoms and signs of an aborted and developing ectopic pregnancy.

A developing ectopic pregnancy	Interrupted ectopic pregnancy
Symptoms of early pregnancy (irregular bleeding or bleeding , nausea, swelling of the mammary glands , bruising of the vaginal mucosa and cervix , softening of the cervix, slight enlargement of the uterus , increased urination) Abdominal pain	Collapse Pulse is fast and slow (110 beats in 1 minute) Decreased QB Hypovolemia Sharp pain in the abdomen and groin Percussion hoarseness and abdominal swelling together indicate the presence of fluid in the abdominal cavity. Signs of irritation of the peritoneum, pallor

Differential diagnosis

a differential diagnosis of ectopic pregnancy with threatening abortion . In other cases, it is carried out with acute or chronic inflammatory diseases of the pelvis, ovarian cyst (rupture or torsion) and acute appendicitis.

UE helps to differentiate between threatened abortion and cystic pedicle torsion.

Quick Actions :

- blood test and preparation for rapid laparotomy . Do not delay the start of the operation until the result of the blood analysis .
- surgery , inspect both ovaries and fallopian tubes :

- if a severe damage to the tubes is detected, perform a salpygectomy (the remaining tubes and the fetal remains are removed together). In most cases, this depends on the choice of the type of operation.

- in rare cases, if the tubes are not damaged, perform a salpingostomy (remains of the fetus should be removed and the tube should be preserved). This should be done only if the preservation of the tubes is important for the preservation of the woman's childbearing function, because the risk of recurrence of the development of ectopic pregnancy remains high.

Autohemotransfusion.

If the bleeding is profuse, for autohemotransfusion to be performed, the blood spilled into the abdomen must be fresh and uninfected (in the late stages of pregnancy, the blood may contain amniotic fluid and other substances and is not suitable for autohemotransfusion). Spilled blood can be taken before surgery or after opening the abdomen:

- When a woman lies on the surgical table before the operation and the abdomen is filled with blood, sometimes the blood can be collected through the abdominal wall into a special system;

- Otherwise open the abdomen:

- collect blood in a sterile container and pass it through gauze to remove clots;

- clean the blood transfusion device with an antiseptic solution and open it with a sterile scalpel ;

- place the woman's blood in the device and carry out hemotransfusion using a filter as usual ;

- if there is no standard bag/vial with anticoagulant , then add 10 ml of sodium citrate to every 90 ml of blood .

Taking the woman later

- Provide the woman with information and advice on maintaining her reproductive function until she leaves the hospital.

- Information about family planning issues is very important considering the risk of recurrence of ectopic pregnancy.

- Correct anemia with iron preparations for 6 months.

- Schedule a female visit in 4 weeks

1.11. El Bugoz

It is characterized by excessive proliferation of chorionic villi

Quick Actions:

- if there is no suspicion of thrush, clean the uterine cavity;

- if cervical dilatation is harmful, use paracervical anesthesia;

- apply vacuum aspiration.

Manual vacuum aspiration is safer and less bleeding. There is a high risk of perforation of the uterine wall when the uterine cavity is scraped with a metal curette;

- three syringes should be ready during manipulation

Due to the abundance of accumulated tissue in the uterine cavity, it is important to empty it quickly: 20 ED oxytocin per 1 l of solution is dripped intravenously (in physiological solution or Ringer-Lactate solution) at a rate of 60 drops per 1 minute to prevent bleeding during manipulation.

Taking the woman later.

- Recommend a hormonal method of family planning to avoid pregnancy for at least 1 year. If the number of children is sufficient and the woman does not plan to give birth again, it is possible to recommend optional enrichment of the fallopian tubes.

- Take a pregnancy test every 8 weeks for a year because of the risk of persistent trophoblastic disease and choriocarcinoma. If the pregnancy test is not negative after 8 weeks and remains positive again during the year, the woman should be referred to a higher level medical center for further observation and treatment.

Table .

1. 12. Family planning methods after abortion

Methods of contraception i	Need to start
Hormonal (pills , injections , implants)	Immediately
condom	Immediately
BIV	Immediately an infection is either known or suspected, delay this method until complete recovery if Nv is less than 7 g/l , delay this method until complete recovery from anemia Recommend temporary methods (eg condoms)
Optional dressing of tubes (sterilization)	Immediately an infection is either known or suspected, delay this method until complete recovery if Nv is less than 7 g/l , delay this method until complete recovery from anemia Recommend temporary methods (eg condoms)

Also , identify other types of reproductive health services a woman may need . For example, some women need: tetanus prophylaxis , treatment for sexually transmitted diseases, cervical cancer screening

CHAPTER II.

2.1. Bleeding in the second half of pregnancy.

Bleeding in the second half of pregnancy is bleeding after 22 weeks of pregnancy. Most often, the reasons can be premature migration of the normally located placenta, incorrect location of the placenta, or the location of the cervix in the center, side, coast, partial, central full, and uterine rupture.

2. 2. EARLY MIGRATION OF A NORMALLY LOCATED PLACENTA

During normal pregnancy and childbirth, the placenta continues to cling to the wall of the upper segment of the uterus until the third stage of labor. When the fetus is born, the uterus shrinks and the placenta moves due to the decrease in pressure inside. Sometimes, in pathological processes, the normally located placenta moves until the birth of the fetus, and this situation is extremely dangerous for the mother and the fetus.

Prolapse of the placenta is more common during pregnancy and the opening of the cervix, and in the second period of labor, placental prolapse is rare. It is a serious obstetric complication and a leading cause of maternal and child mortality. In Uzbekistan, this complication occurs 10 times more often than in warm climate regions. Premature displacement of a normally located placenta leads to death in 2-8% of cases.

Reasons.

In diseases that continue with changes in the vascular system, the placenta often migrates prematurely. Premature displacement of the normally located placenta often occurs when degenerative and inflammatory processes are observed in the uterus and placenta. The walls of the placental vessels of the uterus become thinner, brittle, their permeability increases, they are easily disintegrated, and it is considered a predisposing factor to placental migration.

Late toxicoses of pregnancy, kidney diseases, hypertension
Changes in placental vessels as a result of stroke, heart defects, anemia and other diseases cause the placenta to burn. Due to blood burning between the uterine wall and the placenta, the placenta moves from its attachment point. Inflammation of the uterus after artificial abortion and in the period of chilla predisposes to premature migration of the placenta.

If the umbilical cord is short and the fetal membrane ruptures late, the placenta may be displaced. In a twin pregnancy, placental abruption may occur after the first fetus is born. Rare causes: injuries, neurological factors.

The pathogenesis of the premature migration of the normally located placenta is explained by the rupture of blood vessels, which leads to the violation of inter-villi blood circulation and the formation of a retroplacental hematoma with bleeding. The incidence of this complication is 1.0 to 5%.

The changes that occur during early placental abruption vary widely, depending on whether the placenta is partially or completely displaced. When the placenta is fully displaced, the deposit accumulates between the placenta and the uterine wall, and sometimes the deposit is absorbed into the uterine walls. In this case, the surface of the uterus becomes dark-reddish as a result of burns to the muscle and serous cavity. This change is called Cuveler's uterus, and the uterus loses its ability to shrink in the first hours after birth, eventually losing a lot of weight.

of early placental abruption according to severity: mild, moderate, severe. Depending on the partial or complete migration of the placenta, the clinical picture can be observed in 3 different levels. It's mild. Found in 60% of patients, 15% of the surface of such a placenta has moved, the general condition of the woman has not changed, pulse, blood pressure, fetal heart rate are normal. There is little bleeding from the genitals, sometimes there is no discharge.

Different types of displacement of a normally located placenta: partially from the center, peripheral, completely separated from the center. The moderate level of severity is found in 20% of patients, up to 40% of the surface of the placenta is displaced. A pregnant woman complains of constant pain in the cornea, general malaise, dizziness, nausea. The patient's color is pale, he sweats cold, blood pressure is reduced by 10-15 mm Hg, pulse is accelerated. When palpating the abdominal walls, it is observed that the uterine wall is tense, and the placental branch of the uterus is bulging. The fetus's heart beats faster, sometimes it becomes inaudible. The severe level occurs in 20% of patients, and more than 40% of the surface of the placenta moves. The severe level of placental migration occurs suddenly, the woman's condition worsens, she feels severe pain in all areas of the cornea, fainting, blood pressure drops, pulse slows down, cold sweats. , the colors will be pale. When palpating the uterus, the uterus becomes tense, and due to severe pain, it is not possible to palpate the parts of the fetus. The fetus dies in the womb. Due to the passage of large quantities of thromboplastin bodies through the blood vessels of the uterine wall, the coagulation properties of the blood are reduced (hypofibrinogenemia) or the complete loss of blood (afbrinogenemia) leads to further bleeding in a woman. A woman's life is in danger. In severe cases of premature placental abruption, the urinary function of the kidney is impaired, sometimes completely stopped (ARF), which is referred to as acute renal failure. It should be compared and not confused with placenta previa, uterine rupture, rupture of the oscozoan ulcer, acute cholecystitis, pancreatitis, and appendicitis.

Risk groups.

1. Pregnancy complications: hypertensive syndrome, preeclampsia.

2. Pregnant women with cardiovascular diseases.

3. Kidney diseases and pregnancy.

4. Blood diseases: congenital and acquired coagulopathy.

During the birth process:

1. Very strong labor activity.

2. Discoordination of labor.

3. Unreasonable strengthening of labor activity.

4. Short umbilical system

5. A sharp decrease in intrauterine pressure, in abundance.

Diagnostics.

1. Assessment of the general condition of the woman

2. Assessment of the condition of the fetus

3. UE - detection of changes in the placenta and prevention of complications allows you to get

Lead tactics.

It also depends on the condition of the woman and the fetus and the degree of placental migration. Regardless of the clinical course, tee DICS syndrome or hypovolemia may develop as a result of depletion.

The sequence of measures that should be taken in case of premature migration of a normally located placenta .

1. Evaluation of patient complaints.

2. Correct assessment of general condition.

3. Accurate determination of hemodynamic indicators

4. External obstetric examination:

• assessment of the condition of the uterus (tonus, tension, bulging, pain ricility).

• assessment of the condition of the fetus

5. Take to hospital as soon as possible

Examination through the skin in the hospital:

• to identify labor activity

• to determine the tactics of conclusion and conduct: caesarean section or if the cervix is open, amniotomy and natural vaginal delivery the window

Treatment of premature placental abruption depends on its clinical level and the degree of opening of the cervix during childbirth. If a small part of the placenta has already moved, and the condition of the woman and the fetus have not changed, the woman is immediately admitted to the delivery room, a calm environment is created, and then measures are taken to prevent coagulopathic bleeding (contrical, dicinon, ATF, vitamin E, calcium gluconate, etc.) .k.) From spasmolytic drugs, 2 ml of decoction, 2 ml of papaverine between the muscles, 5 ml of baralgin into the vein are used to stop the contraction of the uterus. Of course, it is also recommended to administer drugs against fetal hypoxia. Amniotomy is performed when the cervix is opened 3-4 cm. This slows or stops placental migration. If the placenta continues to migrate, the delivery is terminated by quick surgery, even if the fetus is dead. During the treatment of a pregnant woman, it is necessary to monitor her pulse, blood pressure, general condition, periodically measure the circumference of the abdomen, the height of the uterine fundus, and listen to the heartbeat of the fetus. In addition, it is necessary to check the composition of the woman's blood and urine, blood clotting properties (coagulogram). If, during these observations, when palpating the uterus, a lump, tension, or pain is felt somewhere, the woman's pulse accelerates, blood pressure drops, and the fetal heart rate changes, all this indicates that the placenta is moving forward. Such a woman needs urgent help. In this case, if the pregnant woman is not in labor, a caesarean section is performed immediately. In the case of Kuveler's uterus, which is burned between the muscle fibers of the uterus at the time of surgery, the body of the uterus, sometimes the entire uterus, has to be cut in order to prevent hypotonic and atonic and coagulopathic bleeding.

In the second period of labor, in order to speed up the delivery, the fetus is delivered by pulling it through the obstetrician's uterus. In summary, moderate to severe placental abruption is treated with prompt surgery if the woman is observed and treated with medication without surgery. Regardless of what measures were used during the delivery of the woman, after the birth of the placenta, a light anesthesia is given, the uterus is examined with a lake, drugs that ensure the contraction of the uterus (methylergometrine) and at the same time blood and its replacement fluids (stabizol, refortan) are dripped. Consequences of frequent displacement of a normally located placenta. Changes that occur in the body of a pregnant woman when the placenta moves are related to the amount of deposits she lost. it depends on the speed of the aid provided, the condition of the body, and the sooner a woman is admitted to the maternity hospital, the more serious complications will occur, as well as the death of mothers and children. If the help is not provided in a timely manner, partial (hypotonic) and complete (atonic) failure of the uterus, decrease in coagulation properties can be observed. Premature displacement of the normally located placenta is a dangerous complication, as a result of which: (DICS) TQI syndrome, Kuveler's uterus, hypo- and atonic bleeding develop.

Treatment. Give birth quickly and diligently. Abdominal caesarean section is appropriate in this case, and in the case of uterus - placental apoplexy - Kuveler's uterus - uterine amputation is necessary. If labor occurs at the end of the first period or premature migration of the normally located placenta in the second period, the labor can be terminated by natural birth. In such pregnant women, the principle of rapid emptying of the uterus is followed. Depending on the obstetric situation, obstetrical forceps are used or fetal segmentation surgery is performed.

The main preventive measures include: kidney and cardiovascular diseases, timely treatment of hypertensive conditions, prevention of abortion and others. At the same time, in order to ensure the proper development of angiogenesis during placentation, it is necessary to remember that the normal course of the menstrual cycle, i.e., the release of estrogens and progesterone in sufficient quantities in the female body, is carried out through a healthy lifestyle and healthy diet. This serious pregnancy and delivery complication can be prevented if serious attention is paid to the period of preparation for pregnancy.

2.3.   LYING IN FRONT OF PLATSENTA

Placenta previa is implantation of the placenta in the lower segment of the uterus before the part of the fetus lying in front of it . Normally, the placenta is located in the body of the uterus, far from the cervix. It should be at least 7 cm high. The placement of the placenta usually takes place at the place where the fertilized egg cell is implanted, because the zygote finds favorable conditions for its development in the uterine cavity and then settles. If the placenta is located on the internal horn, it will cause severe bleeding in the mother, and the risk of severe hypoxia in the fetus will increase. There are many classifications of placenta previa, mainly based on its location relative to the internal cervix. But, most of the time, this relationship leads to significant changes before the onset of labor, so it is better to take into account the time at the beginning of labor.

Variants of pathological adhesion of the placenta

The accepted classification is shown in Table 1.

1 Table .

Classification of placenta previa

Degree	Location
I	Placenta is located below : the placenta is implanted in the lower segment , but does not reach the internal throat .
	Coastal location of the placenta - the tissue of the placenta touches the internal throat a little. The edge of the placenta can be palpated during vaginal examination.
III	Partial placement of the placenta (approximately 30%) is covered by the placenta .
IV	The placenta lies completely in front (approximately - 40%) - the internal throat is closed by the placenta. If the placenta is mainly implanted around the internal cervix , then the term "central location" of the placenta is appropriate .

When the placenta lies anteriorly, it should be taken into account that in women it may become firmly attached or attached. In this case, a pathological adhesion of the placenta to the wall of the uterus is formed. In this case, there is no decidua basalis. The placenta may be indirectly attached to the myometrium (accreta), embedded in the myometrium (increta), or grown into the myometrial layer (percreta).

In the third trimester of pregnancy, placenta previa occurs in 0.1 to 1% according to various authors. Maternal mortality in this case reaches 0.9%. This problem is mainly for multiparous women: its frequency in first-time mothers is 1 in 1500 births, and in the fifth birth, its risk increases to 1 in 20 births!

Placenta previa increases 12 times in subsequent pregnancies compared to women without this pathology.

At the beginning of the second trimester, placenta previa can reach 5% (women who underwent amniocentesis for genetic tests were identified in the group). In 90% of them, placental abruption disappeared by itself. It is interesting to note that women with loss of placenta previa at the end of pregnancy have higher rates of pregnancy complications such as massive postpartum hemorrhage, premature placental abruption, fetal growth retardation, premature births, and perinatal mortality. These complications occurred in 45% of women in whom placenta previa resolved. The pregnancy itself was full of complications. Therefore, such pregnant women are included in the high risk group.

After the placenta is located below it, there are two scenarios of its development - centripetal (in which the placenta lies completely in front) and unidirectional - to the richly vascularized area of the uterine fundus. Such an increase can cause the placenta to lie in front, because the placenta is closer to the bottom of the uterus, and there are not so many blood vessels in the lower segment. Additional evidence of such growth is eccentric, bordered and generally membranous adhesion of the navel. The place where the umbilical cord is attached shows the previous place where the placenta was attached. If the placenta is detected in the fundus of the uterus in the second trimester, it will not migrate anywhere else when the pregnancy reaches term, and the placenta will not lie anteriorly in this woman.

Etiology

placenta previa are abortions, endometritis, history of multiple births, and maternal age (advanced). One of the important risk factors is a violation of the vascularization of the decidual layer due to atrophic or inflammatory changes. The relationship between the number of births and placenta previa can be considered as an etiological factor of this complication. The endometrium may have changes in the implantation site left over from the previous pregnancy, so it is possible that the fetal egg will implant in the lower segment in the next pregnancy.

Also, a correlation between cesarean section and placenta previa and adhesion was found. If there is no scar in the uterus, placenta previa occurs in 0.26%. According to the number of scars, it increases in a linear relationship, if a woman has experienced 4 or more caesareans, this number reaches 10%. If the cesarean operation was carried out due to placenta previa, the pathology is 24%, and if the number of operations is 4 or more, it reaches 67%. A poorly developed decidual layer in the lower segment is the reason for improper adhesion of the placenta, which significantly increases the risk of its true adhesion.

2.4. Clinic of placenta previa

The main symptom of placenta previa is vaginal bleeding, but it does not always appear at the end of the second trimester and at the beginning of the third trimester. This bleeding is not accompanied by a painful syndrome, and usually occurs at rest (often during sleep). If bleeding occurs in third-trimester fetuses, they should be treated as patients with placenta previa, unless other causes are identified. Most often, bleeding begins at 34 weeks of pregnancy, half of patients stay in hospitals until 36 weeks. Only 2% of women with placenta previa experience bleeding before 38 weeks of pregnancy. But in approximately 30% of cases, the first bleeding can occur before the 30th week of pregnancy. Bleeding is not always painless. About 25% of women go into labor after bleeding, in which the membranes may remain intact or rupture. In ¼ of the patients, placenta previa is detected during childbirth, and in these women, the placenta is located only on the coast.

Bleeding is a rupture of the placenta and decidual membrane. The lower segment of the uterus does not contract well, so it cannot compress the spiral arteries. Because of this, the bleeding continues after the birth of the child and after the separation of the placenta. In general, bleeding can be caused by vaginal penetration, sexual intercourse, vaginal trauma, or the onset of labor. The earlier the bleeding starts, the more likely it is that the placenta will block the cervix.
Obstetric tactics depend on the location of the placenta anterior or posterior, the strength of bleeding, the presence or absence of fetal distress, the duration of pregnancy, the presence of the fetus and the degree of anterior lying.

Diagnostics

of placenta previa is made based on complaints, anamnesis, and previa. The anterior segment is usually high because the lower segment is occupied by the placenta . Often there is a malposition of the fetus, usually lying with the bUEocks or transversely (in 33% of cases). If the fetal head comes forward, then the head is located above the level of the small vagina, it is difficult to palpate, especially if the placenta is located on the front wall of the uterus. Clinical data are unequivocal, important, but the final final diagnosis can only be made with the help of ultrasound examination.

If it is used abdominally, then the woman's bladder should be full. In this case, the accuracy of the method reaches 93 to 97%. The remaining cases remain undiagnosed due to the difficulty of viewing the lateral walls of the uterus. Therefore, if the placenta is suspected to be in the front wall of the uterus, then the woman should be offered to empty the bladder, because a mistake can be made when the bladder is full.

Other reasons for incorrect diagnosis are the presence of fibroids in the uterine cavity, contraction of the myometrium, blood clots, late migration of the placenta, the placenta is located behind, when the fetal head passes its edge.

Ultrasound examination using vaginal probes is common in pregnant women with suspected placental abruption. Although it theoretically causes damage to the placenta with the sensor, it is safe. The use of thin sensors allows for better visualization of the lower segment and its relationship with the latent. The sensitivity of the method approaches 100% in these cases, but false-positive or false-negative results can also be obtained, so it is better to use both an abdominal and a vaginal sensor (first with the first, then with the second). Magnetic resonance imaging (MRI) can also be used to detect placenta previa, as this method does not involve any radiation exposure. MRI provides good visualization of the lower segment of the uterus. In particular, magnetic tomograms create a view of placental tissue. But this method is not considered routine, because it is quite expensive, and it is advisable to use it when UE cannot provide information. Unquestionably, this method is indicated when a woman is admitted in childbirth, when the bleeding is not strong. Intra-abdominal examination should be carried out only in the conditions of the operating room, together with a team ready to perform a quick delivery! If there is a slight suspicion that the placenta is lying in front, it is necessary to stop the examination and start cUEing immediately.

2. 5. When the placenta lies in front obstetric tactics

Most women with placental abruption are under medical supervision from the onset of painless vaginal bleeding at the end of the second and third trimesters of pregnancy. An event in any massive bleeding is to stabilize hemodynamics. Therefore, pregnant women with continuous bleeding should be placed in an emergency hospital. It is necessary to start infusion therapy during transportation. Immediately, a large-diameter catheter is inserted into the vein through the puncture of a thick needle, the blood group is determined and instructions are given for hemotransfusion. In the pre-hospital stage, infusion therapy should be started and fetal heart rate should be monitored.

After the mother's condition is stabilized, an ultrasound examination can be performed to confirm the diagnosis. Vaginal examination should not be performed until the diagnosis of placenta previa is ruled out. Most women these days have had an ultrasound scan at least once, so it's worth looking at the results.

A diagnosis of placenta previa requires a caesarean section, but there are exceptions. For example, in the case of placental abruption (grade I or II), after amniotomy, the fetal head can be expected to stop bleeding by squeezing the blood vessels of the placenta. If the bleeding is up to 300 ml, and the above conditions are present, then the pregnancy can be delayed up to 39 weeks and a planned delivery can be carried out naturally using amniotomy and medical stimulation. If the placenta lies more in front (III or IV degree), then the woman is subjected to urgent surgery without waiting for the development of massive bleeding and hemorrhagic shock. Of course, if there is no bleeding, it is better to carry out the examination in a planned way, because urgent surgical intervention has a negative effect on the fetus and increases the possibility of anesthesiological complications. The development of anemia in babies after emergency caesarean section is 27.7%. At the same time, anemia was recorded in only 2.9% of children in the planned practice.

Many authors recommend performing an amniocentesis (if possible) when the full placenta is present and the gestation period is 36-38 weeks, and cesarean section is planned when the information on the maturity of the fetus is obtained (determining the ratio of lecithin, sphingomyelin).

If the pregnant woman does not agree to amniocentesis, then there is no bleeding, then cesarean section can be postponed until 38-39 weeks. The rate of perinatal mortality in placenta previa is primarily determined by gestational age at delivery and, to a lesser extent, by the severity of bleeding and the presence of arterial hypotension in the mother. If the bleeding has stopped on its own up to 3000 ml, then in this situation it is possible to wait, and there is a possibility that the birth will happen by itself. Over the last 50 years, maternal mortality has decreased from 25% to 1%, and infant mortality has decreased from 60% to 10% over the past 50 years as a result of well-thought-out obstetrical tactics.

Over the past 40 years, amazing changes have occurred in the treatment of ectopic pregnancy. Currently, a bleeding woman is not rushed to surgery, even if arterial hypotension develops and is corrected by crystalloid and colloid transfusion. Hemotransfusion is indicated in such cases when the hematocrit is less than 28%, because the oxygenation of maternal and fetal tissues deteriorates at low hematocrit values. In the remaining cases, infusion therapy with hydroxyethylated starch (refortan) can be carried out.

When deciding whether to perform a cesarean section before 36 weeks of gestation, a decision must be made whether pediatric services can help the child or whether the woman should be transferred to a facility that can care for profoundly premature infants. What is important in this situation is not the duration of pregnancy, but the maturity of the fetus.

Tocolytics

The available data on the use of tocolytics in placenta previa are conflicting. In pregnant women with placenta previa, labor may begin up to 37 weeks due to rupture of the membranes or other reasons. The use of tocolytics often lubricates the clinical picture, causing tachycardia in the mother, which is mistaken for a hypovolemic state. The presence or absence of hypovolemia has a profound effect on obstetrical tactics, some obstetricians presuming that placental abruption has occurred. In the same way, other doctors consider bleeding as a sign of placenta previa and not early placental migration, and require the use of tocolytics according to the protocol of the institution and based on the condition of each patient.

Compared to β-adrenomimetics, magnesium sulfate is safer, the mother and child suffer more from bleeding. Hypermagnesemia causes the body's response to maternal blood loss. On the other hand, there are many risks in the use of β-adrenomimetics, for continued use it is better to use magnesium sulfate.

2.6. Tactics of placenta previa

The method of cUEing the abdominal wall depends on the clinical situation . If a woman has previously experienced several surgical interventions, a longitudinal incision is appropriate. The incision of the uterus is resolved during the surgical procedure. If there is a well-formed lower segment, when the placenta is located on the posterior or anterior wall of the uterus, or when the fetus is in a longitudinal position, then a transverse incision can be used. The presence of the placenta in the front wall of the uterus and its grafting in the area of the lower segment are not indications against transverse cUEing. In this situation, the surgeon has two options :

1. Taking the fetus from the placenta a

2. Cut from the top of the placenta .

Both methods carry the risk of serious bleeding if the surgeon tries too hard to remove the fetus without sufficient experience . But it should be remembered that long-term manipulations of a cut or separated placenta can lead to hypoxia and even death of the fetus. In many cases, when the fetus is lying in front, because the placenta is located along the front wall of the uterus, the estimated amount of bleeding during surgery may not be less than 1500 ml, so everything should be ready for hemotransfusion in the operating room. Equipment for autotransfusion is not justified for intraoperative use, because the presence of a large amount of thromboplastin in the collected blood can provoke the syndrome of blood clotting in small blood vessels (DICS). Do not forget about the possibility of an embolism of the jugular waters.

Heavy bleeding may occur in the postpartum period due to poor contraction characteristics of the lower segment. Bleeding from placental adhesions is common and can be stopped by applying hemostatic sutures and manual compression of the placenta. If the bleeding continues, then the uterine arteries and hypogastric arteries are enriched from both sides using a surgical approach. If the measures indicated above are ineffective, it is necessary to perform a hysterectomy.

One of the main risks of placenta previa, especially in women with a history of caesarean section, is placental ingrowth.

Placental ingrowth, especially severe forms - when the chorionic villi grow into the wall of the uterus - is a serious complication of placenta previa.

It is known that five out of six deaths from obstetric hemorrhage occur in cases of placenta accreta. In such a situation, when the surgeon sees a growth, he must act quickly and decisively. A hysterectomy is almost always the only solution. It will be a "delay is death" situation. A lot of bleeding occurs, a large amount of blood transfusion is required, coagulopathy and DICS develop, and as a result, many organ failure is almost impossible to cope with such blood loss.

If the patient is bleeding, then these women have a very high risk of blood loss during the operation and for the following reasons:

1. the placenta may be located in the anterior wall and the obstetrician-gynecologist may accidentally cut it

2. After the birth of the fetus, the elongated lower segment contracts poorly.

3. If there is a history of caesarean section, there is a high risk of placental ingrowth.

In all three cases, profuse bleeding begins.

The delivery of the fetus and placenta almost always stops the bleeding. If it continues , from the atonic lower segment , then oxytocin, methyl gomethrin or prostaglandins are used . Based on data on changes in arterial, central venous pressure and diuresis rate, CBP it is possible to draw a conclusion about bringing it to the norm . Immediately after birth, the newborn requires intensive care, because asphyxia, acidosis, and hypovolemia are most likely.

DICS- Disseminated intravascular coagulation syndrome (TTIQIS) is an example of consumptive coagulopathy, which usually accompanies premature placental abruption, but its presentation is extremely rare. Massive and unnecessary blood transfusions are usually complicated by dilutional coagulopathy.

2.7. Placenta ingrowth

If a pregnant woman has other risk factors that indicate the possibility of placenta previa or placenta accreta, after confirming the diagnosis, the only way to solve the problem is planned cesarean section, followed by hysterectomy.

If placenta accreta is not diagnosed before delivery, then the diagnosis is usually not distinguished in the third period, when the delivery of the placenta is delayed, and even during manual examination of the uterine cavity. An attempt to forcibly separate the placenta is accompanied by fatal bleeding. Blood loss is at least 2 liters.

Large doses of erythrocyte mass, (SZP) fresh frozen plasma and plasma replacement solutions should be transfused, even if the diagnosis is made in time, if a planned cesarean section is performed, then a hysterectomy is performed. The average infusion volume is 3.5 liters.

Placenta percreta is the most severe form of this complication, in which bleeding can occur intraperitoneally or even into the bladder. This condition is life-threatening and is often clinically similar to placental abruption or uterine rupture.

The correct diagnosis is usually made after the birth of the fetus and after attempts to manually separate the placenta and remove it from the uterus, but this remains unsuccessful. All this occurs against the background of heavy uterine bleeding, shock or acute uterine inversion. The choice of treatment method is hysterectomy, but in this case, the patient has a high probability of complications and death. Maternal mortality varies from 67% with conservative treatment to 2% with timely surgical intervention, antibiotics, blood transfusion and fluid therapy.

The risk of developing hemorrhagic shock, sepsis and DICS-syndrome is extremely high in these patients.

Elective caesarean section after diagnosis of placental abruption

Preparation of the patient consists of catheterization of two vessels with large-diameter catheters, insertion of the catheter into the artery, into the central vein, preparation of devices for warming the patient, and at least four doses of blood corresponding to one group. In addition, the blood transfusion service should be informed that a large number of red blood cell masses and fresh frozen plasma suitable for the same blood group may be required at the same time.

The first reason is insufficient pain relief, since hysterectomy after caesarean section is twice as expensive as conventional surgery.

Secondly, the size of the operation is larger, the manipulations on the peritoneum are more extensive, there are more incisions, so strong anesthesia is required.

Thirdly, the uterine vessels thicken and twist, by the end of pregnancy their walls become thinner, so there is not far from the risk of heavy bleeding with any careless action of the operator.

Careful implementation of hemostasis depends on the obstetric team, because the risk of developing disseminated intravascular coagulation (DICS) is very high, and later there may be septic complications.

After the operation, the patient should be transferred to the intensive care unit. It is very necessary to periodically check the hematocrit, the state of hemodynamics, the rate of urination, measure the temperature and monitor breathing.

Blood vessels lie in front

Anterior vascularity is uncommon and is associated with a very high perinatal mortality.

Anterior vascularization is the attachment of the membrane to the umbilical cord, which is located in front of the part where the vessels are located at the lower pole of the fetal bladder.

The veins pass through the membranes and are located between the front part of the fetus and the cervix. They are not protected by anything, which means that they are prone to injury when the pain starts. At the same time, of course, the risk of losing the child is very high, because the diagnosis is made too late for any intervention. Clinically, the bleeding from the genital tract is manifested immediately after the rupture of the membranes and changes in the fetal heart rate. The only solution is to perform an immediate caesarean section. Only this will save the child's life.

2.8. Uterus rupture.

Rupture of the uterus is one of the most serious injuries during childbirth. Violation of the integrity of the uterine tissue during childbirth or crushing of the uterine tissue and the formation of fistulous ulcers as a result.

The most common site of uterine rupture is the lower uterine segment. In rare cases, cracks occur in the area of the domes. Fissures in the body and fundus of the uterus are often observed after caesarean section scars or perforation during abortion. When the fetus is located transversely, it is separated from the domes of the uterus. The break of the domes is near the neck (the wall is relatively thin in this area).

The size of uterine rupture varies. There are two types of cracks: complete and incomplete. Complete rupture is observed in the areas where the serous layer of the uterus is densely combined with other layers, while in the case of incomplete rupture, the mucous and muscular layer of the uterus is ruptured, and the serous part remains intact. Such a rupture creates a hematoma under the peritoneum. Incorrect rupture does not completely occupy the depth of the uterine wall. Sometimes small cracks are observed, in which the injury can be observed in the mucosa or serous layer. Complete rupture of the uterus is observed 10 times more often than incomplete. P. of uterine rupture. L. According to Persianinov (1954). classification.

The lower part of the uterus, the wall is the thinnest part, and this part is often torn. Etiopathogenesis: there are two different theories of uterine rupture: Bandl's and Verbov's theories .

According to the mechanical theory (Bandl's theory), there must be a disproportion between the anterior part of the fetus and the groin, a mechanical obstacle (narrow groin, wrong location of the fetus, too fast labor activity) for childbirth. According to Bundl's theory, when the fetal head encounters an obstruction (anatomical or functional narrow pelvis), it pushes into the pelvic inlet and compresses the cervix against the pelvic bones. As a result, the lower part of the uterus is excessively stretched and stretched, becoming extremely thin. This indicates that there is a risk of uterine rupture. If the woman is not helped in time, the uterus will rupture as a result

Ya.T. According to Verbov's theory, in the absence of pathomorphological changes in the musculature of the uterus, healthy tissue is not torn, but the wall of the uterus with pathomorphological changes is ruptured. Scars on the wall of the uterus, especially after previous operations (caesarean section, after removal of uterine tumors, etc.), previous inflammation of the walls of the uterus, frequent and frequent replacement of the muscles in the uterus of women with multiple births into connective tissue that does not stretch well can lead to spontaneous tearing of the uterine wall. reason will be

Rupture of the uterus under the influence of external forces is mainly caused by excessive stretching of the lower segment as a result of external pressure during childbirth.

Such tearing can sometimes occur in improperly performed operations.

Classification of uterine rupture

1. According to the period of rupture:

A) During pregnancy B) During childbirth

2.   Pathogenetic sign according to A) By itself cracking

- Mechanical (during childbirth and mechanical obstruction in a healthy uterine wall)

Histopathic (pathological changes in the uterine wall when)

- Mechanogistopathic (mechanical   obstruction   to   the pathologically changed uterine wall   effect)

B) As a result of tension: (under the influence of external forces):

- Traumatic (childbirth or unprepared labor in the lower segment of the uterus during pregnancy, rough and improper go)

Mixed: the lower segment of the uterus is elongated and external factors effect.

Clinical course according to

Dangerous

Started

Roy gave

The nature of the injury according to

Crushed

Incorrect

Full

Instead according to:

At the bottom of the uterus

In the body of the uterus

Lower uterus in the segment

From the domes of the womb interruption

Uterine rupture has not yet been fully studied, but it should be said that rupture can occur due to a combination of degenerative changes in the uterine muscle and mechanical effects.

Uterine rupture is divided into risk of uterine rupture, incomplete and full rupture depending on the clinical course.

What are the danger signs of uterine rupture? A pregnant woman is extremely restless and complains of painful cramps, which do not stop even between contractions. The pain gets worse and worse. The woman throws herself in all directions and even tries to stand up, the pain is so great that she does not touch her stomach. Dry tongue and lips. Stroke accelerates.

When examined from the outside, the uterus is elongated and tense, the bottom rises high and is pinned to the subcostal arch. A curved border (contraction) ring with saddle-like pain, usually in front of or above the navel, can be detected through the abdominal wall. This causes the uterus to be in the shape of an "8". will come

At the same time, very tense, painful round ligaments of the uterus can also be identified. When palpating the uterus, severe pain is observed or not heard, especially at the bottom. If timely help is not provided, the wall of the uterus may rupture or tear. In this case, the doctor will see you under deep anesthesia very carefully need

If a little blood-mixed fluid from the vagina is added to the above-mentioned symptoms, it indicates that the uterus has begun to tear. In this case, only the mucous and muscular layer of the uterine wall may be ruptured, the serous layer may remain intact, and blood may accumulate under it and the serous layer may become swollen. At this level of uterine rupture, if a woman in labor is not given emergency aid, i.e., by means of pain relief (under light anesthesia), and if the operation is not performed quickly, the uterus may rupture completely (in which all three layers of the uterine wall are torn), there may be a lot of bleeding in the abdominal cavity, and shock may occur as a result. In this case, the woman passes out, feels nauseous and vomits, breaks out in a cold sweat, has an increased heart rate, her condition worsens, and her face pale his eyes sink into him. It cannot be detected because the blood pressure is low . All these indicate that the situation of the woman is extremely difficult. When holding the abdomen, it is revealed that the fetus is lying under the wall, and the body of the uterus has contracted and hardened on one side of it. In this case, there may be no bleeding or very little bleeding. But all the signs of internal bleeding are clearly visible in a woman. If the midwife can distinguish the above-mentioned signs, it is difficult to determine the risk of uterine rupture in time not.

The uterus can be torn without the above signs or with only some of these signs. This is caused by pathomorphological changes in the walls of the uterus (as mentioned above) or postoperative scars possible

If a woman gives birth within a year or two after a cesarean section, the risk of uterine rupture is greater, as this area can become thinner during childbirth (M. A. Repina 1984). .

When palpating the wall of the uterus, especially the area of the old scar through the abdominal wall, it is determined that some part of the scar in the uterus is thinned and painful. If the uterus begins to tear, this area becomes thinner; this "pit" can be carefully identified with the tip of the scissors (simptom "tip" Kogan A.A.). When the wall of the uterus is torn, blood flows between its wide strands and a hematoma is formed. In this case, as mentioned above, internal bleeding is observed in the woman, and if urgent help is not provided at this time, the woman may die.

Women with a risk of rupture of the uterus should be registered in a special "risk" group at the family polyclinic, paramedic-obstetric centers, at the expense of the local doctor, undergo a medical examination 4-5 times a month, and go to the maternity hospital 2-3 weeks before giving birth. This group includes women with a narrow pelvis, a fetus that may pass through term, a loose abdominal wall, multiple and frequent births, previous surgery (caesarean section), removal of a tumor from the uterus, and others.

provided when there is a risk of uterine rupture during childbirth ? In such moments, giving the woman deep anesthesia, immediately stopping the pain and depending on her condition, one of the following operations will be:

a) all types of cUEing the fetus into pieces; b) caesarean section.

If there are signs that the uterus has begun to tear, only a caesarean section is performed.

When there is a risk of uterine rupture, it is strictly forbidden to put clamps on the fetus's head or turn the fetus to the side of the leg. will be done.

When a caesarean section is performed, the uterus is sometimes removed, but often the uterus is preserved and the tear is sutured. possible

When there is a risk of uterine rupture, especially premature or complete rupture, the woman cannot be sent to another maternity ward or hospital, but a specialist doctor is called to the maternity ward where the woman is lying, and the necessary assistance is provided there.

How and when to operate, whether to preserve the uterus depends on the general condition of the woman, the degree of rupture, age, and obstetric anamnesis. In the prevention of uterine rupture, the correct management of labor and the timely identification of signs of the risk of rupture are of great practical importance. Any tear in the uterus is dangerous for mother and child. The importance of prevention rather than treatment of uterine rupture big

Risk of uterine rupture.

This is a condition that precedes uterine rupture, which is caused by excessive stretching of the lower segment of the uterus and the neck. It's basically a groin It can occur in early labor when the sizes and parts of the fetus do not match each other. The uterus is elongated, the bottom is pulled to one side, the contractile ring remains in the umbilical region, and the uterus takes the shape of an hourglass. The upper part is strongly shortened, the contours are sharply defined, it is located either under the left or right rib. The lower part is a little softer, wider and has a less defined border will be

The round ligaments of the uterus, especially the left one, are mostly tense and painful. The lower segment of the uterus is painful when palpated, and fetal parts cannot be identified. When viewed through the vagina, the cervix is not detected, the cervix is fully open, and the anterior part of the fetus is above the entrance to the pelvis. Sometimes the birth canal is filled with "birth mucus".

The danger signs of uterine rupture? A pregnant woman is extremely restless and complains of painful cramps , which do not stop between contractions. The pain gets worse and worse . The woman tries to throw her body in all directions, and even tries to stand up, because the pain is so great that she does not touch her stomach. Dry tongue and lips. Stroke accelerates.

When examined from the outside, the uterus is elongated and tense, the bottom is raised, and the bottom of the uterus is curved. A saddle - like pain usually in front of or above the umbilicus, a borderline (contraction) mass behind the abdominal wall. I 'm sorry. At the same time, it is possible to determine the round ligaments of the uterus that are very tense and painful . When palpating the uterus, severe pain is observed or not heard, especially below. If timely help is not provided, the wall of the uterus can be partially or completely torn. The doctor is under deep anesthesia with great care must see.

If the above-mentioned symptoms include a small amount of bloody discharge from the vagina, this indicates that the uterus has begun to tear . In this case, only the mucous and muscular layer of the uterine wall dissolves, the serous layer can be completely covered, and blood can collect under it and the serous layer can be doubled. At this level of uterine rupture , if a woman in labor is not given emergency aid , i.e. , by means of pain relief (under light anesthesia) and if the operation is not performed quickly , the uterus may rupture completely (in which all three layers of the uterine wall are torn) , a lot of bleeding in the abdominal cavity , and as a result, shock may occur. In this, the woman is pleased, nausea, vomiting, cold, sweating , rapid heart rate), condition it gets worse and face pale his eyes sink into him. It cannot be detected because the blood pressure is low . All these indicate that the condition of the woman is extremely difficult . When holding the chorion, it is revealed that the uterus lies under the wall of the fetus , and on one side of it, the body of the uterus has shrunk and hardened . I n such a case, there may be no bleeding or very little bleeding . But all signs of internal bleeding are clearly visible in a woman. If the midwife can distinguish the signs mentioned above , it is difficult to determine the risk of uterine rupture in time not.

The uterus can be torn without the above signs or with only some of these signs. This can be caused by pathomorphological changes in the walls of the uterus (as mentioned above) or postoperative scars.

Birth within a year or two after a cesarean section , the risk of uterine rupture is greater, because this area can be thinned during childbirth (Repina M. A. 1984).

When palpating the wall of the uterus , especially the area of the old scar, it can be seen that some part of the scar in the uterus is thinned and painful . If the uterus begins to tear, this area becomes thinner; slowly make this "hole" with the tip of the nail polish remover to determine possible When the uterine wall is torn, blood flows between its wide strands it is poured and a hematoma is formed. In this case, as mentioned above , internal bleeding is observed in the woman, and if urgent help is not provided at this time , the woman may die . possible

Women with a risk of uterine rupture should be registered in a separate "dangerous" group at family polyclinics, health care centers, district doctors , undergo a medical check-up 4-5 times a month, and definitely go to the maternity ward 2-3 weeks before giving birth . This group includes women with a narrow pelvis, a fetus that may be overdue, a ruptured uterine wall, multiple and frequent births, previous surgery (caesarean section) , removal of a tumor from the uterus . thrown and other women enters.

Risk of uterine rupture during childbirth What emergency care is provided when observed? In such moments , the woman is given deep anesthesia and the pain is immediate stop and depending on the situation one of the following operations is performed:
a) all types of cUEing the fetus into pieces ; b) cut cUEing
If there are signs that the uterus begins to tear , only cut cUEing operation is performed.

When there is a risk of uterine rupture, it is strictly forbidden to apply clamps on the fetus's head or turn the fetus to the side of the uterus.

When a caesarean section is performed, the uterus is sometimes removed, but often the uterus is preserved and the tear can be sutured.

When the uterus is in danger of rupture, especially premature or full in case of rupture, the woman cannot be sent to another maternity ward or hospital , but a specialist doctor is called to the maternity ward where the woman is lying on the ground necessary help is displayed.

How and when to perform the operation , whether to preserve the uterus depends on the general condition of the woman , the degree of rupture, age, Obstetric anamnesis . In the prevention of uterine rupture, the correct management of childbirth and the timely detection of signs of the risk of rupture are of great practical importance . Any tear in the uterus is dangerous for the mother and child . It is important to prevent the cure of uterine rupture have

CHAPTER III. 3. 1. HEMORRHAGIC SHOCK

In modern obstetrics, the term "hemorrhagic shock" is defined as a condition characterized by a sharp decrease in circulating blood volume, cardiac output, and tissue perfusion caused by the decompensation of protective mechanisms associated with acute and massive bleeding during pregnancy, childbirth, and the postpartum period.

The development of shock is usually caused by hemorrhages greater than 1000 ml, that is, more than 20% of CBP/OTSK (or a loss of 15 ml/kg of body weight). Continued blood loss of more than 1500 ml (CBP/OTSK more than 30%) is considered massive and life-threatening. The amount of circulating blood in a woman is on average 6.5% of body weight.

3. 2 . Hemorrhagic shock etiology. In pregnant women, the reasons that cause shock to a woman in childbirth and after childbirth include premature migration of a normally located placenta or bleeding when the placenta lies in front, pregnancy in the cervix or in the place of transition from the cervix to the body, when the uterus ruptures, in the III stage of labor, placental separation and discharge disorders, parts of the placenta are caught in the uterine cavity. stay, hypotonic and atonic bleeding during the first chill.

The risk of obstetric hemorrhages is that they are sudden, large in number, and with unknown outcome.

Pathogenesis. Regardless of the cause of massive bleeding, the main link in the pathogenesis of hemorrhagic shock is the imbalance between the reduced volume of circulating blood and the volume of blood vessels, first manifested by macrocirculation disorders, that is, systemic blood circulation disorders, then microcirculatory disorders appear, and due to this accelerated disorganization of metabolism, enzymatic shifts and proteolysis.

The macrocirculatory system is formed by veins, arteries and the heart. The microcirculation system includes arterioles, venules, capillaries and arteriovenous anastomoses. It is known that about 70% of the total CBV/OTSK is in the veins, 15% in the veins, 12% in the capillaries, 3% in the chambers of the heart.

If the blood loss does not exceed 500 - 700 ml, that is, about 10% of CBP/OTSK, compensatory receptors occur by increasing the tone of venous vessels, which are most sensitive to hypovolemia. At the same time, there are no significant changes in arterial tone, heart rate, and tissue perfusion.

Loss of blood more than these amounts leads to severe hypovolemia, which is a strong stress factor. In order to preserve the hemodynamics of vital organs (primarily, the brain and heart), strong compensatory mechanisms are activated: the tone of the sympathetic nervous system increases, catecholamines, aldosterone, ACTG, antidiuretic hormone, glucocorticoids are released, and the renin-angiotensive system is activated. Due to these mechanisms, there is an increase in the heart rate, a delay in fluid and its addition to the blood flow to the tissues, spasm of peripheral vessels and the opening of arteriovenous shunts. These adaptive mechanisms, which cause the circulation to center, temporarily maintain cardiac output and blood pressure. At the same time, the centralization of blood circulation cannot ensure the long-term vital activity of the female body, because it is carried out due to the violation of peripheral blood flow.

Continuous bleeding leads to the exhaustion of compensatory mechanisms and the release of the liquid part of blood into the interstitial space, the development of blood coagulation, sludge syndrome, and the deepening of microcirculation disorders due to a sharp slowdown in blood flow, which leads to deep tissue hypoxia, acidosis, and other metabolic diseases.

Hypoxia and metabolic acidosis disrupt the work of the "sodium pump", increase osmotic pressure, increase hydration, and cause cell damage. Reduction of tissue perfusion, accumulation of vasoactive metabolites contribute to stagnation of blood in the microcirculation system and disruption of the processes of its flow, blood clot formation. Sequestration of blood occurs, which leads to a further decrease in CBP/OTSK. Acute deficiency of CBP/OTSK disrupts the blood supply of vital organs. Decreased coronary blood flow, heart failure develops. Such pathophysiological changes (including blood coagulation disorders and the development of DICS) indicate the severity of hemorrhagic shock.

The level and time of the effect of compensatory mechanisms, the severity of the pathophysiological consequences of a large amount of blood loss depend on many factors, including the rate of blood loss and the initial state of the woman's body. Slowly developing hypovolemia does not cause even significant, fatal hemodynamic disturbances, although it poses a potential risk for an irreversible condition. Small repeated bleeding over a long period of time can be compensated by the body. But compensation very quickly leads to deep and irreversible changes in tissues and organs.

the placenta lies forward .

In modern obstetrics, OPG-pre - eclampsia , somatic diseases (cardiovascular, liver and kidney), anemia of pregnant women, obesity, exhaustion of women during long-term labor, especially if labor pain is accompanied by pain syndrome , sufficient anesthesia with surgical assistance. special importance is given to the cases of non-existence . In such cases, initial conditions for the development of shock, widespread spasm of blood vessels , hypercoagulation, anemia, hypo- and dysproteinemia, and hyperlipidemia can be created. The nature of the obstetric pathology that led to bleeding plays an important role in the development of shock.

3. 3. Clinical presentation of hemorrhagic shock . It is customary to distinguish the following stages of hemorrhagic shock:

I - compensated shock:

II stage - decompensated recurrent shock;

III - irreversible shock.

comprehensive evaluation of the clinical picture of blood loss corresponding to pathophysiological changes in organs and tissues .

Hemorrhagic shock *Stage I* (small withdrawal syndrome or compensated shock) usually develops with blood loss corresponding to 20% of CBP/OTSK (15 - 25% or 700 - *1200 ml of blood loss)*. Compensation for the loss of CBP/OTSK is carried out by the overproduction of catecholamines.

The clinical picture is dominated by symptoms indicating changes in cardiovascular activity of a functional nature: pale skin, destruction of subcutaneous veins in the hands, moderate tachycardia up to 100 beats/minute, moderate oliguria, and venous hypotension. Arterial hypotension is absent or mild. If the bleeding has stopped, the compensated phase of shock can last for a long time. With continuous blood loss, circulatory disorders deepen and the next stage of shock begins.

The second stage of hemorrhagic shock (decompensated reversible shock) develops with blood loss corresponding to an average of 30-35% (25 - 40% or 1200 - 2000 ml) of CBP/OTSK. At this stage, circulatory disorders deepen. Due to vasospasm, the high peripheral resistance cannot compensate for the low cardiac output, so there is a drop in blood pressure. The blood supply to the brain, heart, liver, kidney, lungs, intestines is disturbed, and as a result, tissue hypoxia and a mixed form of acidosis, which requires correction, develop. In the clinical picture, in addition to systolic blood pressure falling below 100 mmHg, strong tachycardia (120 - 130 beats/min), shortness of breath, acrocyanosis against the background of pale skin, cold sweat, restlessness, oliguria below 30 ml/s, heart sounds spasticity, central venous pressure decrease (TSVD) occurred.

Shok 's *III stage* (decompensated irreversible shock)

CBP/OTSK develops with blood loss equal to 50% (40 - 60%, which exceeds 2000 ml). Its development is determined by the subsequent disturbance of microcirculation: capillarostasis, loss of plasma, aggregation of blood cells, extreme deterioration of organ perfusion and increased metabolic acidosis. Systolic blood pressure falls below 60 mm above the wire. The pulse accelerates to 140 rpm and above. Violation of external breathing increases, excessive paleness or marbling of the skin, cold sweat, sudden cooling of the limbs, anuria, stupor, loss of consciousness are noted.

In obstetric practice, the clinical presentation of hemorrhagic shock, in addition to the general symptoms characteristic of this shock, has its own characteristics due to the pathology that caused the bleeding.

Hemorrhagic shock with placenta previa is characterized by severe hypovolemia associated with the background of its development: arterial hypotension, hypochromic anemia, a decrease in physiological growth of CBP/OTSK towards the end of pregnancy. 25% of women develop mild thrombocytopenia, hypofibrinogenemia, and a mild degree of disseminated intravascular coagulation syndrome with increased fibrinolytic activity.

In the shock caused by hypotonic bleeding in the postpartum period, after a short period of unstable compensation, there is a rapidly irreversible state with disseminated intravascular coagulation syndrome with persistent hemodynamic disturbances, respiratory failure, and profuse bleeding caused by the consumption of blood coagulation factors. will come

Premature separation of the normally located placenta usually develops against the background of long-term OPG-gestosis (this is a chronic form of disseminated vascular coagulation, with the presence of hypovolemia) and chronic vascular spasm. Hemorrhagic shock in this pathology is often accompanied by anuria, cerebral edema, respiratory disorders, and a decrease in fibrinolysis.

Symptoms of hypovolemia and external respiratory failure are characteristic of the clinical picture of uterine rupture shock. The development of DICS-syndrome is not so much.

3.4. Diagnosis of hemorrhagic shock .

Hemorrhagic shock is usually diagnosed quickly without difficulty, especially if there is massive bleeding. But early diagnosis of compensated shock, which guarantees the success of the treatment, sometimes causes difficulties due to underestimation of existing symptoms. The severity of shock cannot be assessed solely by blood pressure or the number of recorded blood losses. The adequacy of hemodynamics should be assessed based on the analysis of a complex of very simple symptoms and indicators: 1) skin color and temperature, especially the characteristics of the skin of the limbs;

2) assessment of heart rate;

3) blood pressure measurement;

4) evaluation of the "shock" index;

5) determination of hourly diuresis;

6) CVP/TSVD size;

7) determination of hematocrit indicators;

8) characteristics of the acid-alkaline state of blood.

Peripheral blood flow can be assessed by *skin color* and *temperature* . Warm and rosy skin, pink around the nails indicate good peripheral blood flow, even with low blood pressure. Cold-colored skin with normal and even multiple high blood pressure indicates centralization of blood circulation and impaired peripheral blood flow. Marbling of the skin and acrocyanosis indicate a deep violation of peripheral blood circulation, paresis of blood vessels, this condition is approaching irreversibility.

Heart rate is a simple and important indicator of the patient's condition, compared only with other symptoms. Thus, tachycardia can indicate hypovolemia and acute heart failure. These cases can be distinguished by measuring CVP/TSVD. Blood pressure should be assessed from such positions.

A simple and very informative indicator of the degree of hypovolemia in hemorrhagic shock is the "shock" index - the ratio of heart rate per minute to the magnitude of systolic blood pressure. In healthy people, this index is 0.5; It increases by 1.0 with a decrease of CBP/OTSK to 20-30%; 30 — 50% CBP/OTSK is equal to 1.5 when lost. If the "shock" index is equal to 1.0, the patient's condition may be in serious danger, but if it reaches 1.5, the woman's life is in great danger.

Hourly diuresis is an important indicator describing blood flow in organs. A decrease in diuresis by 30 ml indicates a lack of peripheral blood circulation, and a decrease of 15 ml indicates the irreversibility of decompensated shock.

CVP/TSVD is an important indicator in the comprehensive assessment of the patient's condition. Normal numbers of CVP/TSVD are 50-120 mm above water and can be a criterion for treatment selection. An CVP level below 50 mm Hg indicates definite hypovolemia requiring immediate CBP/OTSK replacement. If blood pressure remains low on the background of infusion therapy, then an increase in CVP above 140 mm of water indicates decompensation of cardiac activity and requires the need for cardiotherapy.

In the same case, low CVP numbers indicate an increase in the rate of volumetric infusion. Combined with the above information, *the hematocrit reading is a good indicator of the adequacy or inadequacy of the body's blood circulation.* Hematocrit in women is 43% (0.43). A decrease in hematocrit below 30% (0.30) is a dangerous sign, and below 25% (0.25) is a sign of severe blood loss. An increase in hematocrit in the III stage of shock indicates that its course is irreversible.

Astrup's micromethod Singgaard-Andersen detection of *KKIX/KOS* is very necessary to get the patient out of shock. It is known that hemorrhagic shock is characterized by the combination of metabolic acidosis with respiratory acidosis. However, alkalosis may develop in the final stage of metabolic disorders.

In modern resuscitation practice, diagnosis and monitoring of the treatment process is carried out under the control of the monitor for the function of the cardiovascular system (macro and microcirculation, osmotic, colloid - oncotic pressure), breathing, urinary excretion and hemostasis system, and metabolic indicators. But at the end of metabolic disorders, alkalosis can develop.

3.5. Treatment of hemorrhagic shock.

To ensure the success of the therapy, it is necessary to combine the efforts of obstetrician-gynecologist - anesthesiologist - resuscitator, and if necessary, involve a hematologist-coagulologist. At the same time, the following rule should be followed: treatment should be started as early as possible, taking into account the cause of bleeding and the health of the woman.

Getting the patient out of shock should be done in parallel with measures to stop bleeding. The surgical volume should be provided with reliable hemostasis. If it is necessary to remove the uterus to stop the bleeding, then it should be done without losing time. In a dangerous situation, surgical intervention for the patient is carried out in 3 stages: 1) laparotomy and stop bleeding; 2) resuscitation measures; 3) continue the operation. The end of surgical intervention for local hemostasis does not mean the simultaneous end of anesthesiological support and artificial lung ventilation (IVL), which are the most important components in ongoing complex shock therapy, which helps to eliminate the mixed form of acidosis.

One of the main methods of treatment of hemorrhagic shock is *infusion-transfusion therapy* and is aimed at: 1) replenishing CBP/OTSK and eliminating hypovolemia; 2) increasing the oxygen volume of the blood; 3) normalization of rheological properties of blood and elimination of microcirculation disorders; 4) correction of biochemical and colloid-osmotic disorders; 5) elimination of acute disorders of blood coagulation.

For the successful implementation of infusion-transfusion therapy to replenish CBP/OTSK and restore tissue perfusion, it is important to take into account the quantitative ratios of infusion agents, volumetric speed and duration of infusion. Taking into account the deposition of blood during shock, the amount of infused fluid should exceed the volume of blood loss: with blood loss equal to 1000 ml — 1.5 times; With a loss equal to 1500 ml - 2 times, with a large blood loss - 2.5 times should be increased. The earlier replacement of blood loss begins, the less fluid is needed to stabilize the condition. Usually, if in the first 1 — 2 hours, 70% of the lost blood volume is replenished, the effect of the treatment will be more favorable.

Based on the assessment of the state of central and peripheral blood circulation, it is possible to determine the required amount of fluid during treatment. Very simple and reliable criteria are skin color and temperature, pulse, blood pressure, "shock" index, CVP/TSVD and hourly diuresis.

The choice of which infusion tools to use depends on many factors: on the initial condition of the pregnant, laboring and postpartum woman, on the causes of bleeding, mainly on the volume of blood loss and the volume of the patient's pathophysiological reaction. They certainly include colloid, crystalline solutions, preserved blood and its components.

Taking into account the great importance of the time factor for the successful treatment of hemorrhagic shock, first of all colloid solutions with very high osmotic and oncotic activity should be used, combining them with crystalloid blood substitutes. Such preparations are HEKs. Keeping fluid in the blood channel, these solutions help to mobilize the compensatory capabilities of the body, so there is time to prepare for the next blood transfusion, which should begin as soon as possible, but all rules and instructions must be followed.

Preserved blood and its components (erythrocyte mass, erythrocyte suspension, frozen washed erythrocytes) remain the most important infusion tools in the treatment of hemorrhagic shock, because at present they can restore the impaired oxygen function of the body. It is permissible to transfuse fresh canned blood. With massive bleeding, the mass of erythrocytes should be 0.5-0.8 the amount of blood loss, but in the course of continuous treatment, due to the risk of developing massive blood transfusion syndrome, more than 3000 ml of blood should not be transfused.

Controlled hemodilution should be combined with the introduction of hemotransfusion of colloids and crystalloids in a ratio of 1:1 or 1:2 to comply with the blood transfusion regimen. This ratio depends on the adaptive characteristics of osmoregulation in pregnant women. For the purpose of hemodilution, any solutions available at the doctor's disposal can be used with quality indicators in any direction. Blood transfusion solutions improve the rheological properties of blood, reduce the aggregation of shaped elements, and thus return the attached blood to active circulation, improve peripheral blood circulation. Preparations made on the basis of HEKs with similar characteristics: infezol, stabizol, refortan.

Adequate treatment of hemorrhagic shock requires not only a large number of infusion agents, but also a significant rate of their introduction (three-dimensional administration rate). In severe hemorrhagic shock, 250-500 ml/min of fluid should be infused per minute. In the II stage of shock, infusion at a rate of 100 — 200 ml/min is required. this speed is solved by catheterization of several peripheral veins or central veins. To save time, it is wise to start the infusion by puncturing the ulnar vein, and then immediately proceed to the catheterization of a large vein, most often the spinal vein. The presence of a catheter in a large vein makes it possible to carry out infusion-transfusion therapy for a long time.

The rate of fluid infusion, the ratio of infusion of blood and blood substitutes, provision of excess fluid should be carried out on the basis of constant monitoring of the general condition of the patient, as well as evaluation of hematocrit, CVP/TSVD, KIH/COS, ECG parameters. Duration of infusion therapy should be strictly individual.

When there is doubt about the severity of hemorrhagic shock, it is recommended to use the following ratio of the components of the infusion: 1 (erythrocytes): 0.2 (albumin): 1 (dextrans): 1 (crystalloids). Stabilization of the patient's condition expressed in the restoration of blood pressure (systolic is not less than 90 mmHg) and satisfactory filling of the pulse, disappearance of shortness of breath, hourly diuresis of at least 30-50 ml and an increase in hematocrit for 0.3, the ratio of 2 blood and fluid may try to drip:1,3: 1. Drip of solutions should be continued for a day or more — until complete stabilization of all hemodynamics.

Metabolic acidosis that occurred before hemorrhagic shock is usually corrected by intravenous administration of 150-200 ml of 4-5% sodium bicarbonate solution.

Addition of 200 — 300 ml of 10% glucose solution with a sufficient amount of insulin, 100 mg of cocarboxylase, vitamins B and C is indicated to improve oxidation-reduction processes.

After eliminating hypovolemia against the background of improving the rheological properties of blood, an important component of normalizing microcirculation is the use of drugs that eliminate spasm of peripheral vessels. The introduction of 2% novocaine solution in the amount of 1 — 1 ml with 20% glucose solution in the ratio of 0.5 or 150: 200 or other infusion means has a good effect. Spasms of peripheral vessels are treated with antispasmodic drugs (papaverine, no-shpa, eufillin) or ganglioblockers of the pentamin type (0.5 — 1 ml of 0.5% isotonic solution of sodium chloride) and heskonia (1 ml of 2.5% solution). can be eliminated by using 150-200 ml of a 10% solution of mannitol is indicated to improve blood circulation in the kidney. If necessary, in addition to osmodiuretics, saluretics are prescribed: 0.04 — 0.06 g of furosemide (Lasix).

Antihistamine drugs that have a positive effect on the exchange process and help improve microcirculation (dimedrol, diprazine, suprastin) must be used. One of the important components of treatment is administration of significant doses of corticosteroids, which improve myocardial function and affect the tone of peripheral blood vessels: initial dose of hydrocortisone — 125 — 250 mg, daily — 1 — 1.5 g. Cardiac drugs are included in the complex therapy of shock after replacing CBP/OTSK (strophantin, corglycon). Significant changes in the blood coagulation system that develop together with hemorrhagic shock should be replaced under the control of a coagulogram. For example, in the I and II stages of shock, an increase in blood coagulation properties is observed. In stage III, coagulopathy may develop due to a sharp decrease in procoagulants and significant activation of fibrinolysis. The use of infusion solutions without coagulation factors and platelets can lead to an accelerated loss of these factors, which are already depleted by bleeding. It can be done by the introduction of non-restored procoagulants ("warm" or "fresh citrated" blood, frozen plasma, antihemophilic plasma, fibrinogen preparations or cryoprecipitate). If there is a need for thrombin neutralization, the anticoagulant - heparin - is used to reduce fibrinolysis - Kontrikal, Gordox.

Time is of the essence in the treatment of hemorrhagic shock. The earlier the treatment is started, the less effort and expense it will take to get the patient out of shock, and the better the immediate and long-term outcome. For example, the therapy of compensated shock is sufficient to replenish the volume of circulating blood and prevent acute renal failure (AE/OPN), and in some cases normalize the acid-base status (KIH/COS). For the treatment of decompensated reversible shock, it is necessary to use all the therapeutic measures. Treatment of stage III shock is often unsuccessful with maximal movement by doctors.

Decompression of a patient with hemorrhagic shock is the first stage of treatment. It is then aimed at eliminating the symptoms of massive bleeding and preventing complications. The actions of the doctor are aimed at the use of vital organs (kidney, heart, liver) and normalization of water-salt and protein metabolism, prevention and treatment of anemia, prevention of infection.

3.6. Organization of emergency assistance. Organizational provision of all stages of emergency care in the hospital is of great importance to reduce maternal mortality from obstetric hemorrhage. Qualified medical care will be successful if the following principles of work organization are met:

1) always ready to help women with major bleeding (blood supply, blood exchange, blood transfusion systems, vascular catheters);

2) the presence of an algorithm of personnel working with large bleeding;

3) constant readiness of the operating system;

4) the possibility of laboratory express diagnosis of the state of vital organs and systems.

Hemorrhagic shock is a condition in which there is a difference between the size of the CBP/OTSK and the size of the vascular channel. This situation causes the following: a decrease in perfusion in vital organs; tissue hypoxia. Without corrective therapy, the following will develop:

- Deficiency of cellular metabolism;
- Enzymatic and metabolic dysfunction occurs;
- Causes cell hypoxia and cell destruction

Macrocirculatory crisis leads to microcirculatory crisis :

Type 1 is observed in non-vital organs: a rapidly increasing decrease in blood flow without stabilization at a subnormal level, that is, there is no compensatory phase in these organs;

2 - type - is observed in vital organs: after 30-60 minutes, a compensation phase occurs in response to blood loss, and the blood flow is set at a subnormal level, until the development of shock enters the terminal stage ;

It changes the rheological properties of blood, leads to the aggregation of shaped e elements and the slad j h phenomenon; tissue ischemia develops .

o the concentration of active polypeptides, acidic metabolites that provide spasm of blood vessels increases;

o sludge syndrome occurs in small vessels and plasma-shaped elements differentiate;

o aggregation of shaped elements

o Conditions for the development of coagulopathy syndrome appear according to the type of DICS.

o other vital organs (liver, kidney, lung, pituitary), the brain and heart are the last to react to this reaction the inside swells);

o is due to the presence of low pressure capillary p ortal system in the vital organs;

o therefore, a decrease in hemodynamics primarily affects these organs;

if not corrected in time , complete injury of macro and micro circulation happens ; /

If hemorrhagic shock develops during pregnancy - it is necessary to eliminate the source and stop the bleeding, if necessary, to resolve the adequate delivery at any period of pregnancy;

If hemorrhagic shock develops in the postpartum period - the 4 "T" rule is used.

Mild hemorrhagic shock: blood loss 1000-1500ml (15-25% of CBV/OTSK).

Moderate hemorrhagic shock: blood loss 1500-2000ml (25-35% of CBV/OTSK).

Severe hemorrhagic shock: blood loss of more than 2000 ml (35-45% of CBV/OTSK).

Take away. Mobilization of all medical personnel, first of all, the responsible physician-anesthesiologist.

Actions : initial assessment of rapid shock: pulse, blood pressure, respiratory rate, body temperature.

* Laying the woman on her left side.

* Warming the woman without heating her.

* Raise your legs.

* Oxygen through mask 10-15l/min.

•	catheterization of peripheral vessels through an angiocatheter or central vein and start infusion .

•	Rapid transfusion of physiological rhythm (1 liter in 15-20 minutes) .

•	The amount of infusion should exceed the estimated amount of blood loss by 3-4 times.

•	Continuation of the started activities and bringing them to the operational block.

•	Order blood (e r. mass, C Z P / FFP, etc.).

OF THE SEVERITY OF ACUTE MASSIVE BLOOD LOSS (body weight 70 kg).

	I	II	III	IV
Volume of blood lost in ml	< 750	750	1500	> 2000
Blood volume lost % CBP/ OTSK	<15	15-30	30-40	> 40
Pulse	< 100	> 100	> 120	> 140
QB	The norm	The norm	decreased	Decreased
Pulse pressure mm. wire he st.	Normal or increased	decreased	decreased	Decreased

Shortness of breath	14-20	20-30	30-40	> 40
Hourly diuresis ml / h oat	> 30	20-30	5-15	no
The state of the central nervous system	Mild excitement	excitement	Indifference	Precoma

30-60 minutes   1 литрinfusion of reosorbilact - 400.0-800.0 ml; FFP / SZP crystalloids 1-2 Colloidal solutions

Additional signs of blood loss : 1. Counting by sight - 30% error from real bleeding . 2. Algover shock index : pulse/ S QB . Norm: 0.5 <0.8 - blood loss < 1000.0 ml. 0.9-1.2- blood loss < 1500.0 ml. 1.2-1.5 - blood loss <2000.0 ml. >1.5 - blood loss >2000.0ml.

3. CVP/ TSVD: +40 - blood loss < 1000.0 ml.

+20 - blood loss 1500.0ml.

<0 - blood loss >2000.0ml.

4. Hematocrit: every 3-4% of blood loss - 500 ml decrease .

5. Weighing the surgical material : 5-7% of its weight - blood loss size

Management tactics for women with hemorrhagic shock are focused on the following: stopping bleeding, ensuring adequate blood oxygenation, restoring CBP/OTSK.

Measures to restore CBP/OTSK:

* Adequate venous access. Access speed and volume. Composition: crystalloids, colloids, blood preparations.

* Cardiomonitor observation, measurement of CVP/TSVD.

* Breathing support.

* Precise control of the injected fluid.

• Bladder catheterization.

* Provide adequate heat regime for the patient (heaters for legs, hands, etc.).

Must remember! The amount of infusion therapy depends on the amount of blood loss and the general condition, the amount of hemodynamics, stopping the bleeding. The amount of transfusion per day should not exceed 20 ml/kg of body weight, otherwise the risk of hypocoagulation may develop.

Restore lost blood

✓ After 4 hours of stopping the blood, a thorough examination is carried out (regular examination every 30 minutes).

✓ the next 24 hours, a personal post will be installed on the woman's bed .

Total volume of infusion

	GradeI bleeding <15% of CBP(OTSK)	II bleeding <25% of CBP(OTSK)	III bleeding <35% of CBP(OTSK)	IV bleeding >35 % of CBP(OTSK)
Volume of blood lost	<1000 ml	< 1500 ml	<2000 ml	>2000 ml
I nfuzi yang i ng volume	150-200%	200-250%	250-300%	300%
Crystalloid/colloids (earth mass+SZP+GEK)	1.5 / 1	2 / 1	2.5 / 1	3.0 / 1
Replacement of body mass (from the volume of lost blood)	-	50%	80%	100%
FFP (SZP)	-	-	50%	50%

CVP (TSVD) - 10- 12 см. over the water

FFP (SZP) - 15-20 ml/ kg weight

Diuresis 30ml/ hour and more

cryoprecipitate 6-8 doses

Fibrinogen dose - 200 mg.

Notes :

- **Crystalloids** are **the** best drugs for hypovolemia correction .

- In order to improve blood transport function , it is recommended to give **erythrocyte mass .**

- Proved koa gulopathy - FFP (**SZP) , cryoprecipitate** casting will be blind .

- to the use of **vasoactive drugs :** the use of inotropes or vasopressors is appropriate after the recovery of CBP (OTSK) .

- If necessary, first of all, inotropic drugs are used , then, if there is no effect, it is better to use vasopressors . However, the use of these drugs can worsen the perfusion and oxygenation of peripheral organs.

- The use of this group of drugs is carried out in resuscitation and intensive care units, where **specialists** the help **of a multidisciplinary team** is needed.

" Tactics of carrying pregnant women with hemorrhagic shock "
hemorrhagic shock occurs in the following cases :

- During pregnancy : **eliminate the source and stop the bleeding** , if necessary, give birth regardless of the period of pregnancy ;

- During Chilla : 4 "T" rule :

- **Read T "** - retention of placental remnants

- **"Tonus"-** uterine atony

- **"Trauma /injury "** - tearing of the uterus or birth canal, inversion of the uterus

- **"Thrombin"** - pre-existing or acquired coagulopathic disorders

On the basis of clinical symptoms, the clinical diagnosis of "shock" is considered :

- Strong bleeding from the vagina / in rare cases, external bleeding is weak or not observed ;

- Pulse accelerated > 110 beats / minute ;

- SA QB < 90 mm.rt.st.;

- The skin is pale, covered with sticky sweat

- Changes in consciousness, sometimes pain in the abdomen can occur .

- Oliguria

- Breathing disorder

Checks:

- General blood analysis (Hb, erythrocytes, platelets)

- General urinalysis

* Coagulogram

* Bedside test i

* EKG

* Determination of the duration of pregnancy

* Bleeding assessment

* Calculating shock index (ShI) according to additional Algover

* Determination of hourly diuresis

* Measurement of MVD (TSVD).

* EKG

Tactics

CALL FOR HELP :

* Call all medical personnel - first of all the doctor in charge and the anesthesiologist

* Immediate primary assessment (Pulse, A QB , respiratory rate , temperature)

* Identify the causes of hemorrhagic shock

* Warming the woman up a bit

* Head down

* liters per minute through the mask .

* 2 peripheral veins angioc a t ether No. 14-16 G with or central venous catheterization

* Start an intravenous infusion using a catheter or a large-caliber needle

* Rapid infusion of physiological solution (1 liter in 15-20 minutes).

* 3 times more than the estimated amount of blood lost during the last shock from bleeding .

* Catheterization of the bladder

* Transfer to the surgical department with continuation of the started activities

* Ordering blood (red cell mass and fresh frozen plasma/ SZP for blood transfusion). Land mass at least 1000ml; SZP - 1000 ml.

Infusion therapy depends on the volume of lost blood and the general condition of the woman, hemodynamics, and the cessation of bleeding .

Kosher treatments and surgeries to stop bleeding

3 steps to stop bleeding :

- Laparotomy:

Ligation of 3 pairs of main uterine blood vessels,

Placing compression stitches on the uterus

Ligation of the internal iliac arteries

Hysterectomy.

- Reanimation measures .

- Continue surgery .

Anesthesiological procedures

- Replacement of lost blood: 3 times the amount of lost blood

- The ratio of crystalloids and colloids is 3:1

- Adequate intravenous fluid infusion is assessed by the woman's hemodynamic status , general condition, and diuresis .

Blood and blood substitutes are given when Hb is less than 70g/l and Ht is less than 25% .

Extricate the patient directly from shock .

- The amount of infusion therapy is decided depending on the degree of hemorrhage and shock . careful monitoring of the woman during the last 4 hours after the bleeding has stopped, constant monitoring of vital organs . During the next 24 hours, a permanent observation individual post will be organized next to the patient's bed in the AvaR department .

Hemorrhagic shock I degree .

- Shock during or after labor and delivery - O75.1

- Bleeding is 1000-1500 ml

- (15-25 % CBP/ OTSK)

- Any kind of obstetrical discharge should be considered as a risk of shock .

Clinical signs

- Pulse is more than 100 beats/minute .

- SA QB 80-90 mm.rt.st.

- The skin is pale
- Consciousness has not changed
- The respiratory rate is slightly increased
- Diuresis 20-30ml/ h

CALL FOR HELP:

- Summoning all available medical personnel - first of all the doctor in charge and the anesthesiologist
- Immediate primary assessment (pulse, heart rate, respiratory rate, body temperature)
- Warming the woman up a bit
- Head down
- liters per minute through the mask .
- 2 peripheral venous angiocatheter No. 14-16 G with or central venous catheterization
- Start an intravenous infusion using a catheter or a large-caliber needle
- Rapid infusion of physiological solution (1 liter in 15-20 minutes).
- It should be 3 times more than the estimated amount of blood lost during the last shock from bleeding .
- Catheterization of the bladder
- Transfer to the surgical department with continuation of the started activities
- Ordering blood (red cell mass and fresh frozen plasma/ SZP for blood transfusion). Land mass at least 1000ml; SZP - 1000 ml.

Hemorrhage and shock II bleeding rate 1500-2000ml (25-35% CBP/ OTSK)

Clinical signs

- Pulse is more than 120 beats /minute .
- SA QB 6 0-8 0 mm.rt.st.
- The skin is pale
- His consciousness is slightly reduced
- Breathing - slight tachypnea
- Diuresis 5 - 20ml / hour

Monitoring:

- Assessment of the condition and monitoring of vital organs
- Assessment of general condition
- Consciousness
- Ps , A QB , body t
- Number of breaths
- Diuresis
- Bloody discharge from the vagina .

CALL FOR HELP:

- Summoning all available medical personnel - first of all the doctor in charge and the anesthesiologist
- Immediate primary assessment (pulse, heart rate, respiratory rate, body temperature)
- Warming the woman up a bit
- Head down
- Restoring the airways
- liters per minute through the mask .
- Infusion therapy: quickly start a drip of 0.9% sodium chloride solution
- Continue anti-shock therapy
- Corticosteroids: Dexamethasone 4-12 mg depending on the severity of the patient's condition
- Hemostatics , antifibrinolytics : Tranexamic acid 1000 mg IV, if necessary, repeated.
- **ATTENTION !** The amount of liquid poured per day **It should not exceed** 20 ml/kg of body weight, otherwise the risk of hypocoagulation may develop .
- If necessary, blood products are injected (erythrocyte mass, fresh frozen plasma , cryoprecipitate)
- Preparation of the operating room

Stop bleeding **during pregnancy** : urgent, regardless of the duration of pregnancy laparotomy, caesarean section.

Ensuring surgical hemostasis **during pregnancy** : laparotomy, ligation of 3 pairs of uterine main blood vessels , placing compression sutures on the uterus , ligation of internal iliac arteries , hysterectomy.

Hemorrhage k shock III degree blood loss 2000ml (35-45% OTSK)

- Pulse is more than 140 beats /minute .

- SA QB 60 mm. wire he st.

- Drowsiness, decreased pain sensation

- Breathing - severe tachypnea

- Diuresis - anuria

- I and continuation of measures in grade II hemorrhage and shock

- Continue filling CBP by infusing crystalloids , colloids , blood components

- Infusion-transfusion therapy: 0.9% sodium chloride solution , HEK solutions , FFP , Erit. mass

- Corticosteroids : Dexamethasone

- Syndrome therapy : cardiac glycosides , cardio- hepatoprotectors , desensitizing agents

- Continue to give oxygen through a mask (10-15l/min) , if necessary, transfer to O'SV

- Adrenomimetics : Dopamine (see appendix)

- Hardware monitoring of hemodynamics :

- Assessment of the condition and control of vital organs, assessment of the general condition, consciousness, Ps, AQB, body t, respiratory rate, diuresis, bloody discharge from the vagina.

Hemorrhage and shock IV degree . Blood loss is more than 45% of CBP

- Pulse more than 140 beats/minute .

- Deep collapse

- Respiratory collapse

- Anuria

- Continuation of measures in I - III degrees of hemorrhage and shock

- Continue filling CBP by infusing crystalloids , colloids , blood components

- Repeated administration of corticosteroids

- Adrenomimetics (dopamine)

- OSV

- C syndrome therapy

Additional treatments

- visual calculation of bleeding - 30% error compared to real bleeding.

Algover shock index (pulse/SAD) :

- <0.8 - blood loss < 1000.0 ml

- 0.9-1.2 – blood loss < 1500.0 ml

- 1.2-1.5 - blood loss < 2000.0 ml

- >1.5 - blood loss >2000.0ml

Central venous pressure (CVP)

- +40 - blood loss < 1000.0 ml

- +20 – blood loss 1500.0ml

- <0 - blood loss >2000.0ml

Hematocrit: every 3-4% decrease - 500 ml of bleeding

Measuring the weight of surgical material - 57 % of body weight - bleeding volume

Management of a woman with hemorrhagic shock is aimed at :

1. Providing blood with adequate oxygenation

2. stop bleeding (see protocol)

3. Restoration of CBP

Arrangements for the restoration of CBP :

Calculation of CBP : body weight x 7.7% (in obesity , body weight more than 100 kg , calculation against ideal body weight - 90 kg).

1. Adequate venous access - central vein or two peripheral veins with a diameter of G more than 16-20 , woman's position - Trendelenburg position

2. Cardiomonitor monitoring, measuring AQB every 15 minutes until CBP is restored

3. Measurement of CVP in the presence of HF and pulmonary diseases allows for safe resuscitation

4. Respiratory support - provision of moistened oxygen through a nasal catheter (4-6 l/min). Solving the issue of transferring the patient to OSV when oxygenation is inadequate or until decompensation.

5. the speed of pouring solutions (only heated and warm) up to 1 liter in 15-20 minutes

6. providing the patient with sufficient heat (pUEing heat on his feet and hands)

7. When AQB is low, in order to prevent injury with pulmonary edema and excess fluid volume, it is necessary to accurately control the volume of fluid administered (hemodynamic monitoring sheet, collection and counting of vials),

8. Catheterization of the urinary bag - control of diuresis, after the recovery of CBP, some diuresis can be stimulated to improve kidney function.

Calculation of the infusion program :

Crystalloids are the best tool for hypovolemia correction

Restoring blood transport activity - er.mass a transfusion

Proven coagulopathy - SZP, cryoprecipitate transfusion

Use of vasoactive drugs: inotropes and vasopressors are used after recovery of CBP/OTSK. If necessary, first of all, it is advisable to use inotropes, if there is no ointment, vasopressors. When using these drugs, it can cause perfusion and oxygenation of peripheral organs. This group of drugs should be prescribed in the OR , where the help of multidisciplinary team specialists is important .

Total infusion volume table #1:

	Blood loss I degree , <15% CBP	**Blood loss II degree , <25% CBP**	**Blood loss III degree , <35 % CBP**	**Blood loss IV degree , >35 CBP**
Volume of blood loss	<1000 ml	< 1500 ml	<2000 ml	>2000 ml
The size of the	150-200%	200-250%	250-300%	300%

infusion				
Crystalloid / colloids (blood , SZP, GEK)	1.5 / 1	2 / 1	2.5 / 1	3.0 / 1
Blood replacement (50% by weight of blood)	-	50%	80%	100%
SZP	-	-	50%	50%

- HEK agents are infused after hemostasis is achieved , when bleeding is only on vital instructions.
- 1-2 degree bleeding, 60% of the rate of infusion is given in the first 2 hours , and the rest in 4-6 hours .
- 3-4 levels 70% in the first hour , the rest in 4 hours .

Pharmacological support for the cardiovascular system

Agent	O datative quantity	Effect
Inotropic agents		
Dopamine	1-3 µg/kg/min	Enhance diuresis Vasodilation
	2-10 µg/kg/min	Increase of YuUS/ChSS Increased heart rate
	>10 mcg/kg/min	Peripheral vasoconstriction UUS/ChSS and increased myocardial contractility
Dobutamine	2-10 µg/kg/min	UUS/ChSS and increased myocardial contractility Postnagruzk a decrease
Vasopressors		
Phenylephrine (Mesaton)	1-5 µg/kg/min	Peripheral vasoconstriction

| Noradrenaline | 1-4 µg/min | Peripheral vasoconstriction |
| Adrenaline | 1-8µg/min | Peripheral vasoconstriction |

Control : A QB , UQS /ChSS , breathing , laboratory tests (hemoglobin, hematocrit, platelet count , fibrinogen, bedside test) , RaO2

Instructions for transfer to OSV:

1. Hypoxemia (when RaO2<60mm.rt.st. FiO2 >0.5).

2. NS more than 40 minutes

3. Low inspiratory strength .

4. It should be carried out until the patient decompensates . Blood loss of 2-3% of body weight is a signal .

large cuff size and low pressure of endotracheal tubes .

Hemorrhagic shock and resuscitation measures should be taken simultaneously with bleeding.

Treatment effect : ESR/ChSS < 100, SA B is ± 15% of the norm (below 110 mm Hg) , diuresis ≥ 1 ml/kg/ h , hematocrit ≥ 30%

of hemorrhagic shock , a positive result can be achieved when central hemodynamics and tissue circulation are restored during the first 6 hours.

Application :

• Algover shock index i is the ratio of pulse to systole k AQB: PS /SA QB .

• According to Lee-White (bedside test i) blood clotting time determination method

• principle: based on determining the time of blood clot formation in venous blood .

Method : Bedside test - Clotting time of 1 ml of venous blood within 1 minute from the time of collection .

Technique :

Take a capped glass tube and warm it to 37°C, then draw 1 ml of venous blood. Then hold the test tube firmly in your hand and observe the formation of a blood clot every minute. Clot formation: A complete blood clot can be seen by inverting the tube. Coagulopathy is suspected when a blood clot forms slowly (more than 7-8 minutes) or when a soft blood clot that breaks down quickly is formed.

In obstetric practice, hemorrhagic shock can occur with any amount of bleeding .

Surgical interventions in obstetric bleeding

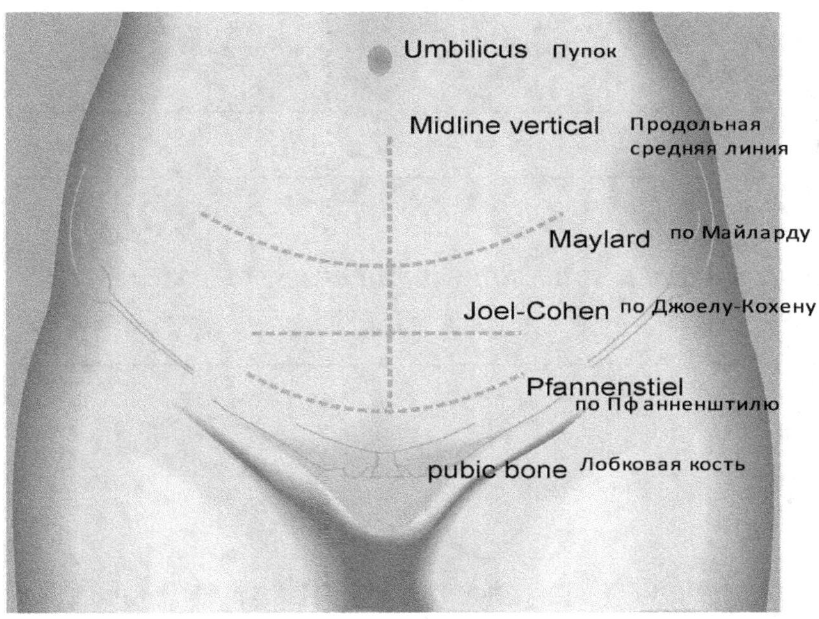

Figure 1. Laparotomy methods

Supravaginal amputation without excess uterus

The position of the patient on the operating table is horizontal. An inferior midline laparotomy, which is acceptable for vision , bypasses the umbilicus on the left . In addition, the operation is performed by Pfanenstiel (Pfannenstiel) from the transverse entrance.

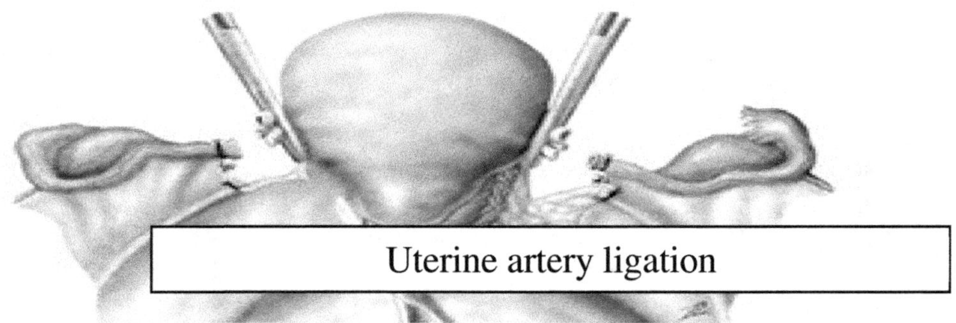

Picture 2 . *Supravaginal amputation without excess uterus*

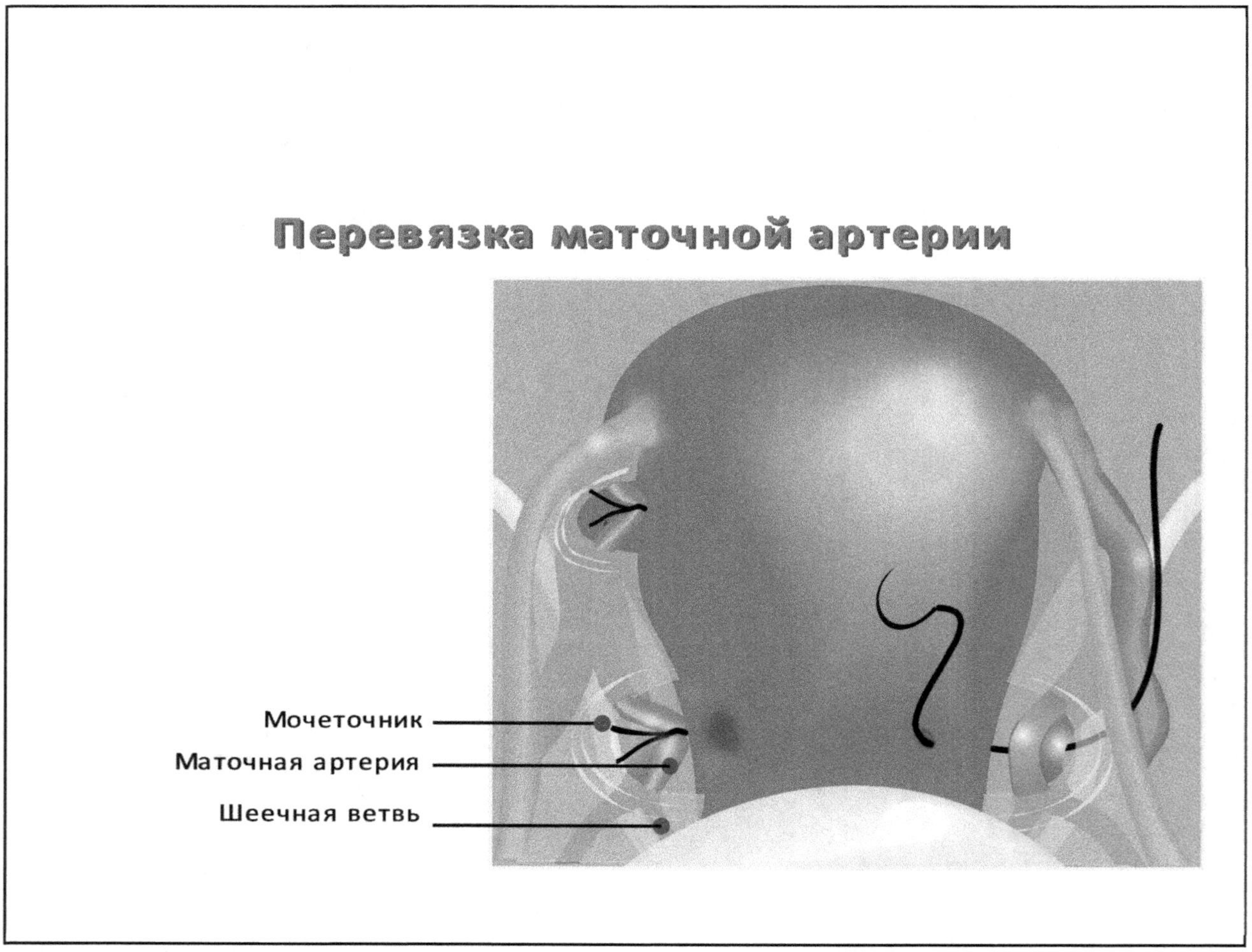

After revision , uterus released . Delimiting the operative field from the intestines, long clamps are placed on the ribs of the uterus and removed to the wound. This leads to the tension of the round ligament of the uterus, which helps in its intersection and ligation. After that, the infundibulum-pelvis ligament is opened, which is also thickened and a ligature is placed. The tips of the forceps should be directed towards the uterus. In the same way, manipulations are performed from the opposite side.

The front and back layers of the abdomen are separated and lowered below the bladder. The back of the abdomen is cut below the internal neck of the uterus. Clamps are applied perpendicular to the side of the uterus at the border of the internal throat on both sides of the fallopian tubes. For hemostatic purposes, the ends of the contraclamps are placed almost close to each other in the corner of these clamps. The vessels of the uterus are cut between the clamps and tied with silk. CUEing the uterus with a scalpel begins with a transition from the posterior wall above the separation of the sacral ligaments to the anterior wall above the vascular culm. Thus, after the removal of the uterus, a cervical clot is formed with the edges of the cervix, which allows the cervical canal to be closed. 3-4 ketgut sutures are put on Kultya . Peritonization with the help of the peritoneum is performed using a continuous catgut suture. Hemostasis is checked and the abdominal cavity is closed layer by layer.

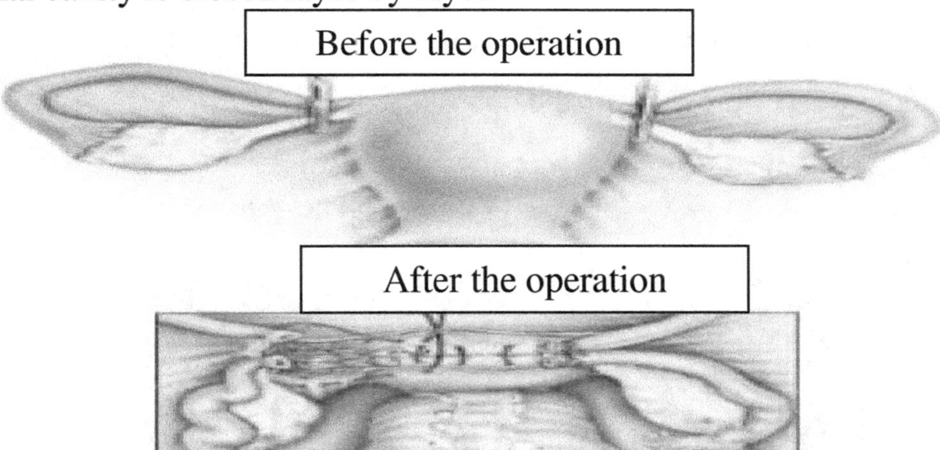

Picture 3 .

a microirrigator is left in the abdominal cavity for the purpose of hemostasis control .

Extirpation of the uterus without excesses

The position of the patient on the operating table is horizontal. An inferior midline laparotomy, which is acceptable for vision , bypasses the umbilicus on the left . In addition, the operation is performed by Pfanenstiel (Pfannenstiel) from the transverse entrance.

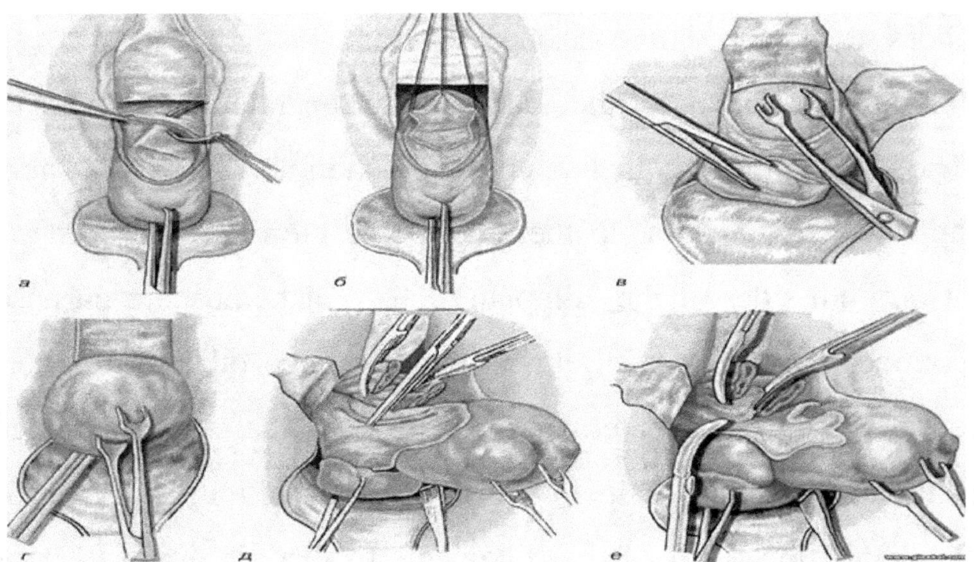

Picture 4 . Extirpation of the uterus without excesses

After revision , uterus released . Delimiting the operative field from the intestines, long clamps are placed on the ribs of the uterus and removed to the wound. This leads to the tension of the round ligament of the uterus, which helps in its intersection and ligation. After that, the infundibulum-pelvis ligament is opened, which is also thickened and a ligature is placed. The tips of the forceps should be directed towards the uterus. In the same way, manipulations are performed from the opposite side.

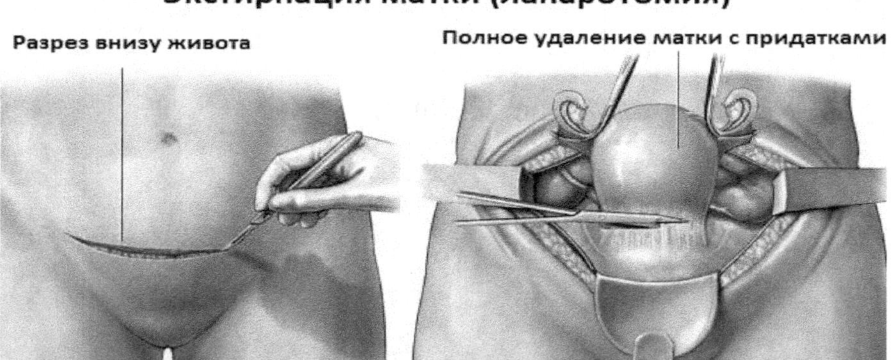

Picture5 . Laparotomy

The front and back layers of the abdomen are separated and lowered below the bladder. The back of the abdomen is cut below the internal neck of the uterus. Clamps are applied perpendicular to the side of the uterus at the border of the internal throat on both sides of the fallopian tubes. For hemostatic purposes, the ends of the contraclamps are placed almost close to each other in the corner of these clamps. The vessels of the uterus are cut between the clamps and tied with silk. CUEing the uterus with a scalpel begins with a transition from the posterior wall above the separation of the sacral ligaments to the anterior wall above the vascular culm. Clamps are placed on the cervical cell directly near the angle of the uterus, the fibers are cut and covered with catgut. The separation continues to the walls of the vagina. At the border with the bladder, the prevaginal dome is opened and pre-marked with Koher's clamp. The uterus is cut at the level of the domes, long clamps are placed on the vagina. Peritonization with the help of the peritoneum is performed using a continuous catgut suture. Hemostasis is checked and the abdominal cavity is closed layer by layer. According to the instructions, if necessary, a microirrigator is left in the abdominal cavity for the purpose of hemostasis control.

TEST:

1. Which of the following tests is not included in the necessary examinations of uncomplicated pregnancy:

a) XG re-examination ;

b) Determining the amount of Nv ;

c) serological examinations ;

d) cytological examination of the cervix ;

e) determination of blood group and rhesus factor

2. Which of the following is not among the most common causes of ectopic pregnancy :

a) endometriosis;

b) chronic salpingitis;

c) adenomyosis;

d) coverage with BIV ;

e) diverticulum of uterine tubes ;

3. Ectopic pregnancy is not recommended :

   a) salpingoectomy or ;

   b) salpingoovarioectomy or ;

   c) longitudinal salpingostomy or ;

   d) resection of the fetal part of the tubes ;

   e) squeeze the fetus from the tubes ;

4. Causes of ectopic pregnancy are not included :

   a) fibrosis of the uterine tubes ;

   b) dysfunction of the ciliated epithelium of the uterine tubes ;

   c) blockage of the distal part of the tubes (hydrosalpinx);

   d) narrowing of the tubes ;

   e) having a false path in the pipes ;

5. How does the amount of XG change in advanced pregnancy :

   a) increases by 2 times ;

   b) will increase by 3 times ;

   c) increases by 4 times ;

   d) does not increase ;

   e) increases by 10 times ;

6. How does the amount of XG change in an underdeveloped pregnancy :

   a) 2 times ;

   b) 3 times ;

   c) 4 times,

   d) Does not increase

   e) 10 times.

7. Possible signs of pregnancy.

 a) blueness of the vaginal mucosa, roughness of the mammary glands, the uterus
     is not enlarged.

 b) roughness of the mammary glands, enlargement of the uterus, blueness of the
     mucous membrane of the vagina and cervix, cessation of menstruation.

c) hardening of the breasts, enlargement of the uterus, pink color of the vagina and cervix.

d) blueness of the mucous membrane of the vagina and cervix, menstruation is smooth.

e) smooth menstruation, hardening of the mammary glands, XG-negative.

8. Clinical signs that are not characteristic of Ellie cow:

a) Menstruation, uterine enlargement does not correspond to the duration of pregnancy.

b) bilateral enlargement of the ovaries,

c) bloody discharge mixed with blisters,

d) multiplying XG by 1000 times,

e) the size of the uterus is equal to the duration of pregnancy.

9. Tactics for carrying fifty cows:

a) scraping the uterine cavity,

b) cUEing ,

c) posterior dome puncture,

d) measuring the uterine cavity with a probe,

e) amputation of the uterus above the vagina,

10. Symptoms not typical for ectopic pregnancy:

a) posterior dome puncture is positive,

b) that the fallopian tubes have changed by means of laparoscopy, blood in the abdominal cavity has been determined,

c) delay of menstruation by 10-15 days, the uterus is larger than normal.

d) AB low, tachycardia.

3.During menstruation, the size of the uterus is normal.

11. Treatment of ectopic pregnancy:

a) scraping the uterine cavity,

b) stitch

c) removal of the uterus above the vagina,

d) uterine extubation,

e) separating the placenta with a finger,

12. Signs of miscarriage:

a) 1. small size of the uterus compared to the period of pregnancy, pain in the lower abdomen, vaginal discharge, closed cervix;

b) the uterus is typical for the period of pregnancy, pain in the lower abdomen, the cervix is closed, the uterus is in tone;

c) typical for the period of uterine pregnancy, pain in the lower abdomen, the cervix is passing 1 finger;

d) typical for the period of uterine pregnancy, pain in the lower abdomen, the cervix is 1 finger long, a lot of bloody discharge;

e) all answers are correct

13. What are the signs of an abortion that has started:

a) small size of the uterus compared to the period of pregnancy, pain in the lower abdomen, vaginal discharge, closed cervix;

b) aching pain in the lower abdomen, typical for the period of uterine pregnancy, the cervix, the examiner passes the fingertip;

c) the size of the uterus compared to the period of pregnancy, pain in the lower abdomen, bloody discharge, the cervix is passed by the examining finger;

d) the uterus is of normal size, bloody discharge is coming, the cervix is closed;

e) the uterus is of normal size, bloody discharge is coming, the examiner passes the finger of the cervix;

14. Symptoms of premature miscarriage:

a) the uterus is of normal size , there is a lot of bloody discharge, the cervix the inspector is swiping ;

b) the size of the uterus is smaller than the period of pregnancy, painful pains in the lower abdomen, a large amount of bloody clot-like discharge , the cervix is not past 1 finger ;

c) the size of the uterus corresponds to the period of pregnancy, pains in the lower abdomen, bloody discharge, rubbing, the cervix is closed ;

d) the uterus is of normal size , the cervix passes 1 finger ;

e) the size of the uterus corresponds to the period of pregnancy , the pain in the lower abdomen , the secretions are heavy, the cervix is closed ;

15. Signs of an abortion on the way :
   a) the size of the uterus corresponds to the period of pregnancy , pain in the lower abdomen , bloody discharge, closed cervix ;
   b) the size of the uterus is small, painful pain in the lower abdomen, bloody discharge in large quantities , the cervix is passing 1 finger, the fetus is in the cervix ;
   c) the size of the uterus corresponds to the period of pregnancy , pain in the lower abdomen , bloody discharge in large quantities, the cervix is closed ;
   d) small, bloody discharges from the uterus, the cervix passes the tip of the finger ;
   e) the size of the uterus is suitable for the period of pregnancy , the pain in the lower abdomen , bloody discharge, the cervix passes the fingertip ;

16. Doctor's tactics in case of complete miscarriage :
   a) all answers are correct ;
   b) anti-inflammatory treatment ;
   c) cleaning the uterine cavity ;
   d) uterine contraction drugs ;
   e) anemia against anemia ;

17 . The risk group of uterine rupture includes:
   a) Pregnant women with uterine scars
   b) Secondary infertility
   c) A woman who gives birth again
   d) ECO
   e) Multiple pregnancy

18. Bleeding after childbirth is determined:
   a) Bleeding more than 0.5% of body weight
   b) Sudden bleeding after childbirth

c) Postpartum bleeding of more than 300 ml

d) Postpartum bleeding of more than 250 ml

e) Bleeding more than 1000 l

19. In case of obstetric bleeding, a transfusion is started to quickly fill the AYUK:

a) Crystalloids (physical solution, Ringer's solution)

b) dextrans (polyglucin, reopoglucin)

c) blood of the same group

d) frozen plasma

e) GEKs

20. Bleeding after childbirth can be caused by tearing of the cervix, vaginal wall, and diaphragm:

a) If the placenta is intact and the uterus is reduced

b) The integrity of the placenta is broken and the uterus is shortened

c) Constellation broken, uterine atony

d) Violation of placental integrity, uterine hypotonia

e) Constellation whole and external bleeding

21. If the loose uterus does not shrink even after the massage of the uterine fundus, the next step is:

a) Administration of uterotonics

b) Bimanual uterus click

c) intravenous fluid infusion

d) Applying ice to the lower abdomen

e) Laparotomy

22. The frequency of the risk of rupture of the uterus in the lower transverse scar of the uterus:

a) less than 1%

b) more than 5%

c) 1%

d) 5%

e) up to 10%

23. What is done first in pathological postpartum bleeding:

   a) see the birth canal,

   b) manual examination of the uterine cavity,

   c) burn the terminal to the parameter

   d) compression of the aorta

   e) Laparotomy

24. Premature migration of a normally located placenta in a pregnant woman with preeclampsia Tactics.

   a) Cesarean section operation

   b) Stimulation of labor with prostaglandins

   c) Use of hypotensive drugs and reopoliglyukin, heparin solution

   d) Administration of magnesium sulfate solution

   e) Protease inhibitor use

25. "Migration" of the placenta occurs because it is the uterus

   a) If it is on the front wall

   b) on the orca wall if b dies

   c) right on the wall if b dies

   d) If it is on the left wall

   e) If at the bottom

26. In the 3rd trimester of pregnancy, when the UE of the uterus is performed, for the diagnosis of the placement of the placenta below, how many centimeters is its tip above the internal throat

   a) 5-6 cm

   b) 11-12 cm

c) 9-10 cm

d) 7-8 cm

e) 3-4 cm

27. What are the types of placenta previa ?

    a) Full, with edge, side

    b) Light, medium

    c) Acute, chronic

    d) Primary, secondary

    e) Central, peripheral

28. Show the reason that causes the placenta to lie in front:

    a) Tr ophoblast nidation function violation

    b) Uterus myoma, chronic pyelonephritis

    c) eclampsia, swelling

    d) Preeclampsia, eclampsia

    e) Uterine inflammation

29. If the placenta is located in any part of the uterus, it is called placenta previa:

    a) In the lower segment of the uterus, partial or complete closure of the uterine cervix

    b) Uterus in the body

    c) On the back wall of the uterus

    d) at the bottom of the uterus

    e) in the front wall of the uterus

30. If early migration of a normally located placenta occurs in the 1st period of labor, which obstetric tactic is used:

    a) Cesarean section surgery

    b) Obstetricians

c) Vacuum extraction

d) natural birth

e) labor induction

31. Cuveller of the uterus which one to the situation thief will come

a) Migraine

b) The arrival of the placenta in front

c) early discharge of sewage

d) uterine rupture

e) seam failure

32. Compared to the classic clinical presentation of the disease one does not enter

a) Fetus heart hit normally

b) Kind of mining separation coming

c) uterine pain

d) in uterine tone

e) onset of pain

33. Premature migration of a normally located placenta is a common cause:

a) Hypertensive conditions

b) Premature discharge of water

c) absolutely tiny navel

d) severe pain

e) uterine prolapse

34. Premature migration of a normally located placenta on the front wall of the uterus is characterized by:

a) local pain

b) hump and backside painful

c) Corinne Swelling on the front wall

d) swelling in the legs

e) fainting

35. Star partly ahead in the location birth understand the tactics why related.

a) to the amount of bleeding

b) to the condition of the cervix

c) to the part where the fetus is located in front

d) to a woman's age

e) to the obstetric situation

36. Manual examination of the uterus after childbirth when the placenta is partially anterior:

a) depending on the amount of blood loss

b) not necessarily

c) depends on the condition of the baby

d) a must

e) regarding instruction

SITUATIONAL TASKS

Example 1

18- year-old patient was sent to the hospital with a complaint: aching pain in the lower abdomen , bloody discharge from the genital tract . 1 abortion in the anamnesis, complicated by endometritis.

The menstrual cycle is not disturbed. Last period 2.5 months ago.

On vaginal examination, the cervix and vaginal mucosa are blue, bloody discharge is average. The cervical canal examiner is passing the fingertip. The uterus is enlarged for 8-9 weeks of pregnancy, in tonus.

Diagnosis: 1. Abortion on the way .

2. Initiated abortion.

3. Ectopic pregnancy .

Tactics. 1. Scraping the uterine cavity .

2. Conservative treatment .

3. L aparatomy.

Example a 2

19 -year-old woman who is pregnant for the first time is 7-8 weeks pregnant

.

Vaginal examination: the cervical canal 3 смis open up to 2nd , the size of the uterus is equal to 5-6 weeks of pregnancy , a lot of bloody discharge .

Diagnosis: 1. Premature abortion

2. Lying in front of the partner

3. El Bugoz .

Tactics. 1. Scraping the uterine cavity .

2. C- section operation

3. Probing the uterus .

Example 3

A 29-year-old patient was diagnosed with secondary infertility. The uterine cavity was scraped 3 times. Fallopian tubes are permeable. Ovulation induction with clomiphene was performed. After 2 months of treatment, there was bloody discharge from the genitals before the expected period. The pains are located in the groin area, irradiating to the shoulder blade and shoulder. KB 70\40 mm. sim.ust. Pulse 112 beats.

Vaginal examination: the uterus is slightly enlarged, mobile. In the area of the uterus excesses are detected, the posterior dome is flattened, painful, irradiating to the posterior urethra.

Diagnosis : 1. ovarian apoplexy

2. Inflammation of the uterus, ectopic pregnancy

3. Abortion on the way

Tactics: 1. hysterosalpingography

2. Lap a rotomy

3. Laparoscopy

Example 4

An 18-year-old patient complained of pulling pains in the lower back and lower back, nausea, and some bloody discharge from the genitals. Anamnesis: Menstruation is regular 4-5 days, last period 2 months ago.

Vaginal examination: the mucous membrane of the vagina is blue, the cervix is formed, the external cervix is tight, the uterus is enlarged equal to an 8-week pregnancy, the uterus is painless, the discharge is medium with blood.

Diagnosis: 1. Initiated miscarriage

2. Menstrual cycle disorder

3. Ectopic pregnancy

Tactics: 1. hormonal treatment

2. Conservative treatment

3. operative treatment

Practical skills for "Bleeding in pregnancy" :

1. Collection of anamnesis ;

2. Determination of presumptive and suspicious signs of pregnancy ;

3. Seeing through the skin ;

4. Seeing the cervix in the mirror ;

5. Assessment of blood loss ;

6. Hysterectomy ;

7. Features of hysterectomy in El Bogozo ;

8. Determination of pregnancy period in UE ;

9. Interpretation of results ;

10. Back dome puncture ;

11. Opening the abdominal wall layer by layer ;

a. salpingoectomy;

small caesarean section ;

12. Cesarean section of the lower segment of the uterus.

13. M agistral blood vessels ;

14.   Uterine amputation operation;

15.   Determination of blood coagulation according to Lee-White ;

16.   Determination of blood coagulation by bedside test ;

17.   Preparation for blood transfusion ;

18.   Venepuncture;

19.   Hemotransfusion;

20.   Creating an infusion-transfusion program.

REFERENCES :

1. Aylamazyan E.K. // Emergency care for extreme conditions in obstetric practice: Guide. / Ed. E.K. Aylamazyan. -L., 2005.Duda I.V., Duda V.I. Clinical obstetrics. - Minsk, 2007.

2 Duda I.V., Duda V.I. Clinical obstetrics. - Minsk, 2007.

3 Integrated management of pregnancy and childbirth. Providing assistance during complicated pregnancy and childbirth. A guide for doctors and midwives. WHO. UNFPA. 2009.Serov V.N., Strizhakov A.N., Markin S.A. Practical obstetrics. - M., 2007.

4 Strizhakov A. N. Clinical lectures on obstetrics and gynecology. M. Medicine 2010.

5 Enkin M, Keirs M, Neilson D i dr. Leadership and effective assistant during pregnancy and childbirth. -SPb., Izd.: "Petropolis". 2003.

6 Primary, perinatal and postnatal care. Educational seminar. WHO. 2003g. - Geneva.

7 The increase in the quality of perinatal assistants and obstetric care facilities of the Republic of Uzbekistan (national standards for the management of pregnancy

and childbirth at various obstetrical facilities). Tashkent 2014

8 V.I. Duda and soavt. "Gynecology". Minsk. 2004.

9 V.I. Duda and soavt. Obstetrics. Minsk. 2004.

10 V. Shevchenko, I. Sidorova "Current issues of obstetric pathology."

11 Management of pregnancy and childbirth during complicated course. Geneva.2005.

12G.M.Savelyeva "Gynecology". Moscow - 2004

13 Integrated management of pregnancy and childbirth. Guide for doctors. Geneva. 2002.

14 K.N. Zhmakin, V.I. Bodyazhina - Emergency care in obstetrics and gynecology, M. 1986.

15 Kulakov A.I. Guide to safe motherhood. Moscow. 2000

16 Kulakov V.I., Serov V.N., Abubakirova A.M., Fedorova T.A. Intensive care in obstetrics and gynecology: (efferent methods). 2008.

17 M. Enkin "Effective obstetric care." St. Petersburg.2001.

18 E.K. Ailamazyan – emergency care for extreme conditions in obstetrics. L. 1985-2000

www.ingramcontent.com/pod-product-compliance
Lightning Source LLC
LaVergne TN
LVHW080222200726
843510LV00007B/1034